Super Science

Unleash Your Superpowers With These Fun Experiments

Cris Johnson

ElementarySchoolAssemblies.com

Super Science: Unleash Your Superpowers With These Fun Experiments

Cris Johnson

Copyright © 2020 Cristopher J Johnson
Cover design by Kerry Watson - Fiverr.com

All Rights Reserved.

All rights reserved. No part of this publication may be reproduced, distributed, or transmitted in any form or by any means, including photocopying, recording, or other electronic or mechanical methods, without the prior written permission from the author, except in the case of brief quotations embodied in critical reviews and certain other non-commercial uses permitted by copyright law.

First Printing: September 2020

ISBN#: 978-1-7356893-0-2

Cris Johnson, Inc.

Event Horizon Publishing

8310 Lockport Rd.

Niagara Falls, NY 14304

Phone: (716) 940-8963

Portions of this book were originally developed as part of a live school assembly experience for elementary school children

Cristopher J. Johnson

Cris Johnson, Inc.
8310 Lockport Rd.
Niagara Falls, NY 14304

(716) 940-8963

www.ElementarySchoolAssemblies.com

Cris Johnson is available to speak at your elementary or middle school on a variety of topics. Call (716) 940-8963 for booking information.

Why Read This Book

Cars, smart phones, televisions, ceiling fans, dishwashers, and on and on...none of it would be available to us without advances in science. Science is amazing by itself, but by combining superheroes and science, this book is an irresistible combination to young readers! Why?

Kids LOVE superheroes! By using simple science experiments that seem to reproduce superhero powers, kids will be hooked! On top of that, readers of this book will be able to get access to videos of each experiment conducted and explained by the author.

With *Super Science*, the lessons and concepts are easy to understand. This book is funny, silly, and features a lot of weird ways to illustrate scientific principles for young readers. From asking the question "Can a hot dog fly?" as a way to explain the Scientific Method to wondering "Did the President of The United States steal my bicycle?" to illustrate what makes a good hypothesis, and why dropping pizza on your sidewalk will help you understand Potential and Kinetic energy, this book is a wild, wacky way to learn fun science experiments that anyone can do safely.

Written by a School Assembly Expert with 20 Years' Experience Holding the Attention of Children

Cris Johnson is an experienced school assembly presenter, having traveled across North America delivering programs. He performs 400 shows a year and offers over a dozen assembly topics including bully prevention, drug awareness, character education, reading, recycling, and many more. Learn more at www.ElementarySchoolAssemblies.com

Cris Johnson is also a Board-Certified Hypnotist & Instructor through the National Guild of Hypnotists and a Master Practitioner & Trainer of Neuro-Linguistic Programming, giving him unique insights into how people of all ages think and process information. Cris is an accomplished speaker on topics such as stress management, teambuilding in the workplace, communication strategies, and more.

Do you want Cris Johnson to be the motivational speaker at your next event?

Call (716) 940-8963 or visit

www.FunCorporateTeamBuilding.com

What others are saying about this book

"This incredible interactive guide to superpowers and science showcases the scientific method. Written in an appealing format, this book releases the creativity, the passion and imagination of the child in an easy to read, easy to understand way. It is a bridge that can only help young minds to fully engage in science while these children are conducting awesome scientific experiments in an amazing, fun-loving way at home using ingredients that are readily available and affordable."

—Robert Fried,
M.Ed. College of New Rochelle, Retired Special Education Teacher

"I can summarize the contents of this book in one word: Awesome! Although this book is geared to young readers, the book is highly readable, engaging, and very entertaining for readers of all ages. I couldn't wait to do some of the experiments myself. Although you may be tempted to jump ahead to the experiments, do read the first few chapters. They lay a lot of the groundwork for what follows. Cris' humorous writing style and simple language appeal to a young audience. The book is fun, informative and demonstrates how we all can have superhero powers."

—Joann Abrahamsen,
former elementary school teacher

"Cris Johnson's *Super Science* book is really neat! He talks to students right at their level, and explains nifty science things in a fun way. I can see all kinds of kids getting excited about this book and learning a lot about the four sciences. There are fun, easy experiments with safe materials and lots to see and do. It's a great way to challenge your brain (and maybe learn a little bit about superheroes, too!)"

—Jim Kleefeld,
M.Ed

"As a former science teacher myself, I think Cris Johnson has got a winner with *Super Science*! The allure of being like a superhero is sure to appeal to students and teachers alike. His 'top 10' approach to describe and guide through each experiment is a wonderful way to make the science easier to understand and follow. Cris' first-person way of writing and talking to the students makes this a fun and an educational book. I love how each chapter ends with 'cool science questions' to further prompt the student-scientist to keep learning. Cris is able to connect to a young reader with ease - 'smelly molecules' and 'flying hotdogs' are guaranteed to excite a junior scientist into diving into this book."

—Peter Bugnet,
former science teacher

"Everyone who works with young scientists will want this publication. It's a truly brilliant resource. This book is an incredible guide to support parents and teachers to tap into the superpowers of science. The author has done the hard work to create meaningful experiences! Do yourself a favor and purchase this valuable resource!"

—Patti Sapp,
M.Ed

"Cris is an experienced and entertaining performer, and that experience shows in his book. The book is funny, relatable, and accessible to everyone. Adults and kids alike will enjoy this adventure into science, experimentation, and imagination. Who knows, maybe you will even get bitten by a radioactive spider! **(My understanding from Cris is that radioactive spiders are not included with the book purchase.)**"

—Adam Gertsacov,
Dadapalooza Blog - dadapalooza.com

Table of Contents

Chapter 1: Introduction: Science & Technology 1

Chapter 2: The Scientific Method: Flying Hot Dogs & Stolen Bicycles 13

Chapter 3: The Four Sciences 25

Chapter 4: X-Ray Vision 37

Chapter 5: Creating Solid Matter Out of Thin Air! 46

Chapter 6: Telekinesis: Stopping a Falling Object 62

Chapter 7: Telekinesis: Make Objects Fly into the Air 75

Chapter 8: Mutations 89

Chapter 9: Creating Evil Science Serpents! 100

Chapter 10: Passing Through Solid Objects! 111

Chapter 11: Invulnerability 123

Chapter 12: Shrinking 133

Chapter 13: Predicting the Future 143

Chapter 14: Controlling the Elements 162

Chapter 15: Conclusion 174

BONUS Chapter for Parents & Teachers 176

Cris Johnson

Chapter 1: Introduction: Science & Technology

Do you think science is boring?

Isn't that a weird introduction to a book about science?

But it's true: When I was in school, I thought science was boring! I hated learning about science because I couldn't imagine how useful this stuff would be in my life... well, all except learning about dinosaurs...dinosaurs are awesome!

It wasn't until I started watching the Discovery Channel and shows like *Mythbusters* that I really began to enjoy science. The experiments they did looked like SO much fun and the best part was they explained all of the scientific principles in ways I could understand. I felt smarter after watching those shows, and I realized I could add to conversations by talking about the science behind some of the things people I talked to noticed every day. It was fun!

Super Science

I should explain my job, that is, what I do when I'm not writing books.

My name is Cris Johnson and I perform school assembly programs.

You know how sometimes a person or a group will come to your school and you'll get to get out of class for a while and sit in a room with the entire school as this visitor talks to you about something important? That's what I do: I get kids out of class!

So in a way, I'm already a superhero, at least to students like you!

I go into schools and talk to kids about really important stuff, like treating each other how you want to be treated, why reading books is so awesome, and stuff like that.

My job is traveling all over the United States doing school assembly programs in schools, but I don't just talk to kids during my shows. I do magic shows! Not only do I do magic shows, but my magic tricks each have a message! That means that while I'm doing a trick, I use that trick to get a point across. Like in my No Bully Zone show, I ask to borrow a kid's one dollar bill. Then I talk about how important it is to ask for help if you need it. Then I change the kid's one dollar bill into a five dollar bill! So the magic makes it easier for the kids watching to remember the important stuff I talk about.

Cris Johnson

Even though I mostly do magic shows, several years ago I decided to start offering schools a science assembly program. I only had one rule: My science show had to have NO magic in it. Zero. Nada.

I didn't want magic in my science show because magic defies science. If you watch one of my shows, you'll see me float a student from the audience in the air! This is very cool to see, but if we are trying to learn about gravity, then showing a magic trick that defies that scientific principle isn't a very good way to learn about science.

Think about it. If we are learning about gravity, which is the force that holds us to the ground and keeps us from floating away, then seeing that rule broken is only going to confuse you...and me. I mean, I'm not a full-time scientist. I'm learning this stuff right along with you in a lot of cases!

So after I performed my first science show for several years, I decided to write a second science show simply because this stuff is so much fun! I wanted this show to be different from the first one, so I decided to combine the science show with something else I love: Superheroes!

I love superheroes because they can do such amazing things and it's so cool that they help people who need help. We might have superhero powers, but all of us could learn how important it is to help other people from superheroes.

So the idea for the Super Science book came from that second assembly program. My idea was that I would take

Super Science

superhero powers, things that don't exist in the real world, and see if I could find science experiments that would reproduce, simulate, or at least sort of look like a superhero power!

Here's the bad news: Unfortunately you won't become a Superhero after learning these experiments, but you CAN become a "Super Scientist" by learning this stuff! Your family, friends, and teachers will be so impressed with what you know!

These are super power science experiments that both boys AND girls can enjoy! I mention that because years ago, there was this bone-headed belief that girls couldn't be interested in science stuff like boys could. If you have ever heard that, it's SILLY and so totally wrong. Science is something that boys and girls can be equally good at doing - it just comes down to how hard you study and do the work.

You might wonder where these experiments came from. I found some of these experiments by reading books. Others I found by watching cool YouTube videos. And some experiments I decided to try because I saw them on *Mythbusters* years and years ago, before most of you reading this were even born! (Yeah, I'm kind of old.)

Whenever possible, for each experiment, I will give its background, the science behind it, and all kinds of details. You're going to learn a lot of cool stuff about science!

Cris Johnson

One rule I made for myself when I wrote this book was to use experiments where you don't need a lot of money. I mean, I saw the *Mythbusters* guys blow up a cement mixer truck once! I don't have the money to blow up a truck…or even a car. Maybe I can blow my nose, but that's about it.

So all the experiments in the book are very affordable. That's important.

Another rule? Every experiment had to use stuff that would be easy to get from a store. In other words, no fancy or dangerous chemicals, nothing with a skull and crossbones on it, and nothing you should have to get off of the Internet. (I mean, if your grocery store is out of something, you CAN get it off of the Internet, but you shouldn't have to.)

I also made a rule that these experiments in the Super Science book were super-easy. I wanted easy experiments that you can do pretty quickly because that's how we learn: we start off learning easy stuff, then we work our way up to things that are harder and harder.

It's sort of like reading…actually, it's exactly like reading! Try to think back when you were really little and you were learning to read the alphabet. Just learning to read one letter at a time. Then you worked your way up to simple words. Then whole sentences. Then paragraphs, then books, and presto! You're now reading a book about science written by some guy you've never heard of.

I also picked easy experiments because at the end of each chapter, I'll challenge you to do the experiment in different ways. Maybe you can change the ingredients to see if something different happens. Maybe you can change the amount of the ingredients to see if something different happens. Or in some cases, I include ways to have contests with your friends! Stuff like that.

Now, can I tell you a secret? When I was a kid in school, I had a hard time learning certain things. I eventually learned that one of the best ways to learn was to get the information in different ways. In other words, if I read about something in a book in class, sometimes that was enough and I'd learn whatever it was I was trying to learn.

Other times, reading wasn't enough. Then maybe I'd ask my teacher to help me and when the teacher used different words to explain it, then I'd get whatever it was I was trying to learn.

I'm going to do my best to do that with this book. See, I'm going to do my best to describe the experiments, what's going to happen, why it's happening, and all of that important stuff as best as I can. And on top of that, this book will have in it a very special webpage where you can see videos of me describing, setting up, and doing all of the experiments in this book!

I have to tell you…many of these experiments are SUPER messy, so you'll not only get to see me making a big ol' mess, but you'll understand why it's best to do many of these experiments outside!

The way the rest of the book is going to be mapped out is pretty much going to be the same for every experiment. I'll list things in this order, in sort of a 'top ten' list, like this:

1. **What is the superhero power?** Here is where I'll list the superhero power we are exploring.
2. **Materials needed:** This is where I'll list the stuff you'll need to gather up. If it's something weird, I'll give you some ideas of where to get it or other stuff you can use instead.
3. **What is the experiment?** This is where I explain what the actual superhero power is that we will be doing plus all of the steps in the experiment.
4. **Why is it a "superhero power?"** For something to be thought of as a 'superhero power,' that means it must either break the rules of science, or at least be a power that is far above what a 'normal' person could do.
5. **Where should I do this?** For safety, I'll tell you where the best place is to do these experiments. The good news is some of these experiments can be done pretty much anywhere! How cool is that?
6. **Safety:** Here is where I'll explain any possible dangers, things that can go wrong, mistakes I made, and stuff like that.
7. **What are some others ways to experiment with this:** Here I'll give variations, different things you

can do with the experiment to change it, make it more fun, have contests with your friends, and more!

8. **The science of the experiment:** After all of the sections involving the actual experiment, I'll get into more of the details of the experiment, the science behind it, the background of the experiment, and other details, such as (in some cases) how long the experiment has been around, and things like that. You'll find that many of the things in this book may seem new and exciting, but the ideas behind them may have been around a long time!

9. **Similar experiments:** Here is where I will list experiments to look for that are kind of like the main experiment we are learning about in each chapter. You may get ideas for more cool experiments...or even an idea to create your own experiment!

10. **Which of the 4 sciences does this fit:** You'll learn in a later chapter about the 4 sciences and why those categories are the ones science experiments fall into and you might find it interesting to find out which one of these four categories the experiment we are working on falls into.

11. **The website where you can see the experiments**: Yup, for every experiment in this book, there's a video of me doing the experiment and explaining the science behind it! (And I even

get my super-cute dog Stanlee involved in one experiment!)

This will actually be the same website repeated each time for every experiment (I have all of the videos on one webpage) but I can imagine that some of you reading this might skip around from one experiment to another, so if you wanted to grab this book one day and skip to the experiment in Chapter 7, you'd be able to see the web address for that experiment right there in that chapter. But I'm hoping you actually decide to read this book all the way through as there's so much cool stuff in every chapter.

Oh, and to access the videos, you will need to ask one of your parents to fill out the contact form on that page.

12. **Cool Science Questions:** Here, at the very end of the chapter, I will ask you some questions about the chapter you just read, so you can see how much cool stuff that you can remember!

Another important thing - I tried really hard to write this book as sort of a conversation we are having instead of me just lecturing. I want this book to have some great information in it but also be fun and kind of goofy...like me!

Oh, this is SUPER important: If your mom or dad or teacher is wondering about this book, you can assure them that the experiments here are very safe. Things CAN go wrong, of

Super Science

course (that's true with ANY science experiment), so this book will explain what could possibly go wrong with each experiment, any mistakes I myself have made learning this stuff, and suggestions for staying super safe!

Also, you have to PROMISE me you'll ONLY do experiments when you have a grownup with you, okay? Pinky promise? I wrote this book because I wanted everyone reading this to have fun with science, but I don't want letters from angry parents telling me that their kid blew themselves up into a million pieces after reading my book.

Speaking of 'letters,' you would probably be amazed at how many letters were mailed when I was your age. Of course, these days you know all about faster ways of sending written messages to people: email, Twitter, Instagram, texting, and probably some new thing that will come out right after this book is published.

Want to know why YOU probably won't have to mail very many letters in your life? Because of SCIENCE!

Science is responsible for all the cool stuff you have in your life. Smart phone? Science. Want to know how?

Simple - a bunch of scientists decided they wanted to work on a way to make it so we could take our phones with us everywhere.

Cris Johnson

I'm sure you have seen all kinds of old TV shows or movies where the people talking to someone used a phone mounted on the wall with a cord. Then maybe your parents told you stories about the first cordless phones where you could go to any room in the house, but you still couldn't drive away while talking to someone.

The thing to remember is that what seems old now was once thought of as really advanced...until the next new thing comes along. Like cell phones.

Work on cell phones actually started back in the 1940s! A bunch of scientists said to themselves, "Boy, it sure would be nice to be able to go outside and stuff and still talk to our friends," so they started working on the first cell phones.

Then cell phones came out and you could walk around and talk to people...but all you could do was talk to people! And those first cell phones available to regular people like you and me were big! My grandfather had one: it had a shoulder strap and it was like three times bigger than my mom's purse!

So what happened next? Scientists began working on making those phones smaller and smaller until they were small enough to fit in your pocket.

Then more scientists started working on phones that could send texts, and it led to these scientists figuring stuff

Super Science

out and pretty soon people could either talk or send text messages on their phones.

After that, more scientists decided to make it so you could play games on your phone, watch videos, all kinds of stuff...all because of science.

Whenever we get new stuff like a game, or iPads, or whatever, it's because of experiments scientists do to figure things out.

The most common way to run experiments and figure this science stuff out is by using what's called the Scientific Method.

In fact, that's our next chapter! See you there.

Cris Johnson

Chapter 2:
The Scientific Method:
Flying Hot Dogs & Stolen Bicycles

Let's look the Scientific Method, because that is going to help you out in science experiments. There are five steps in the Scientific Method:

1. Ask a question
2. Make a guess
3. Do something
4. Observe your results
5. Then you have your answer.

So, the first step is just that - it begins by asking a question.

The question can be anything, as long as you don't know the answer. So a rule in the Scientific Method could be: **The first step is to ask a question.**

Let's pretend you walk outside and notice your bicycle is missing. You remember you left it outside the previous night.

Super Science

The question you ask is, "Hey, where's my bike?"

The next step is to make a guess. Where do you think your bike went?

By the way, if you want to sound like a really smart scientist, call your guess your "hypothesis." That word is just a fancy way of saying "guess."

To look at it another way, a "hypothesis" is using all of the knowledge we have about a subject, even if it isn't very much, to make the best guess we possibly can.

Your guess might be: "I think the President of the United States sent the Secret Service to our house to take my bike."

That's a guess, all right. But is it a good one? Is it a good hypothesis?

Think about it for a minute. The President of the United States is a really powerful person. Would he or she have the ability to send the Secret Service to your house in order to steal your bike?

Yes, the President definitely has that kind of power.

But does that make your guess a good one?

Think about it for a moment.

Cris Johnson

The President is super powerful and super rich. The President makes a lot of money doing a really difficult job. Even after the President is no longer the current President, he or she can still make a lot of money going to schools and companies and talking to groups.

That means the President is super rich and has enough money to buy thousands of bikes just like yours.

Because the President is super rich and can buy any ol' brand new bike he or she wants, is the guess of the President stealing your bike actually a good hypothesis?

Not really.

Back to the morning after your bike is missing: What would be a better guess or hypothesis?

You might say, "I'll bet another kid took my bike!"

Do some people take stuff that doesn't belong to them? Unfortunately, yes, that does happen. That's why houses, cars, and even school lockers have locks on them: to protect peoples' stuff.

So while you don't know for sure whether or not another kid took your bike, it's a better hypothesis than thinking the President of the United States took your bike.

We can make that another rule in the Scientific Method: **Our hypothesis is an educated guess.**

Super Science

How do you find out whether or not your guess is right?

In our missing bike example, let's pretend you go back into your house and you ask your mom where your bike is.

She says, "I put it in the garage last night after you went to bed. You have to remember to take care of your bike! Someone might steal it!"

So now you have the answer to your question.

But here's the deal: You didn't have to do all of the steps in the Scientific Method because your mom gave you the answer.

Here's another question, and it's really weird:

Can hot dogs fly?

Okay, before I continue, I KNOW that is super weird! Everyone knows hot dogs can't fly!

But how did people first find out hot dogs couldn't fly? What if you were from a different planet and you visited earth and you encountered a hot dog, just sitting there, on the sidewalk, waiting for a bus? (And for that matter, if you were from a different planet, you probably wouldn't know what a bus is either, but let's just stick to hot dogs.)

How would you know whether or not a hot dog could fly?

In the Scientific Method, after you ask the question, you make a guess: You might say, "No, I don't think hot dogs can fly."

Just as with the missing bike, this guess is called your **hypothesis**.

Now, if you had never seen a hot dog before, you would only be making a guess. And you know from school, if you make a guess, you might be right, or you might be wrong.

How do you find out whether or not your guess is right? You need to **do something**.

That's right, you have to actually do something to find out if your guess is right or wrong.

What kind of things could you do? Lots of stuff.

You could ask your mom or dad, "Do hot dogs fly?" They might look at you kind of strangely, but you would probably get an answer. Same thing with one of your teachers: you could ask one of them, and they would give you some kind of answer.

Here's the thing, though. We are pretending that you are from another planet and you've never seen a hot dog before. In fact, let's pretend that NO ONE has ever seen a hot dog before!

If no one knows what a hot dog is, then no one knows whether or not it can fly. So you could ask every person

Super Science

you know about this weird tasty-looking meat thing sitting there waiting for a bus (that's right, I didn't forget about that bus, either), but since no one knows the answer, asking people for the answer isn't going to help you answer your question.

So, to find out whether a hot dog can fly means you have to do something else. You have to do something involving the hot dog itself to find out whether or not it can fly.

This is called an "experiment." Scientists like to use fancy-sounding words, but in this case, calling what you are doing an "experiment" makes a lot of sense, because while you are "doing" something, you are carefully deciding what you are going to do in order to get an answer as to whether hot dogs can fly.

What kind of experiment could you do to find out if hot dogs can fly? Well, you could throw that hot dog into the bathtub. That would definitely answer the question of whether or not a hot dog floats, but would doing an experiment of throwing a hot dog into the bathtub full of water answer the question "Can hot dogs fly?" Nope.

So now we know that "doing something" in the Scientific Method means "doing" MORE than just asking someone else the answer. Number one, that's too easy, and number two, scientists are usually trying to find the answers to questions that no one knows the answer to.

We also now know that "doing something" in the Scientific Method should make sense for the question we are trying to answer.

So, we can say this as a rule in The Scientific Method: ***A good experiment has to seem to make sense and hopefully will answer the question that a scientist is trying to answer.***

Back to our flying hot dog question: What would be a good experiment to test whether a hot dog can fly?

Here's a few ideas I had while writing this book:

1. You can throw the hot dog down the stairs.
2. You can throw the hot dog across your yard outside.
3. You could drop your hot dog out your window.

Because those experiments involve trying to get the hot dog into the air, they are all good ideas to try if you had never seen a hot dog before and wanted to find out if they can fly.

(Please don't actually do any of these experiments. You already know that hot dogs can't fly. Besides, if your parents catch you throwing hot dogs out windows, they will write me more of those angry letters.)

So, you do something. You conduct an experiment. Maybe you conduct several experiments. Is that enough?

Super Science

Nope.

You have to watch what happens.

Here's what I mean. Let's pretend you throw a hot dog out the window, but instead of watching what happens, you turn away from the window and start playing video games. Later, you remember the hot dog and you turn back to the window, and you see the hot dog laying on the ground.

Can you say for sure that the hot dog didn't fly?

No, because of this: What if the hot dog DID fly around, and then it got tired and landed on the ground, so when you looked out the window, you just happened to see the hot dog resting after a long flight?

I know what you are thinking: "No, you crazy dude, hot dogs can't fly!"

I know, I know...but remember, we are pretending that no one has ever seen a hot dog before so no one knows.

This means a good scientist can't just do the experiment, not pay attention, and just assume everything turned out like he or she thought it would.

That's step 4 in the Scientific Method: **A good scientist observes the results of the experiment.**

So, when doing an experiment, you have to watch and see what happens.

Cris Johnson

We now come to step 5: your **ANSWER.**

Let's say you have asked your question (step 1), you've made a guess that seems to make sense (step 2), you've done your experiment (step 3), and then you've observed your results (step 4), you have an answer (step 5)...does that mean you are done?

Maybe, but probably not.

I know that may seem like a not-very-good answer, so let me explain.

There are some experiments that answer a question and prove a guess right or wrong so well on the first try that the scientist definitely is sure of the answer. Then there are other experiments where the answer is not as clear.

Here's an example of each.

Let's pretend a scientist wants to know what will happen to a handful of grapes if they are run over by a bus driving at 50 miles per hour (that's equal to a little over 80 kilometers per hour if your country uses the metric system). The scientist's hypothesis might be, "I'll bet that bus is going to squish those grapes completely flat."

So the scientist puts some grapes on the road and then he signals his friend the bus driver to start driving along the road. When the bus gets up to 50 miles per hour and hits the grapes, what happens?

Super Science

SQUISH!

Those grapes are now SO flat.

That's what we would call a "definitive answer."

And the bus driver keeps on driving. He has to get to the bus stop where those hot dogs are waiting (remember?).

Let's look at another example. Let's pretend that Cooties are really a disease. And a scientist is working on a cure. He comes up with a medicine that he thinks will cure Cooties completely. So he tries this medicine on just one person and Presto! The Cooties are gone!

Does that mean the scientist has found the perfect cure for Cooties?

Nope.

With something like trying to cure a disease, there are a lot of things a scientist may not know about that might keep him or her from knowing the answer for sure until after a LOT of tests are done.

Some of these things that might change the answer might be:

- A person's age
- What kind of medicine a person already takes
- Whether the person was sick with other diseases
- What race a person is

- How healthy a person is

There's actually a LOT more things than that, but that gives you an idea of what might prevent just one test giving a scientist a "definitive answer."

So usually when they are testing something, a scientist will run the same experiment over and over, a whole bunch of times.

Then, maybe the scientist may change one little thing about the experiment and run the experiment several more times.

Getting back to our hot dog experiment, let's say a scientist throws 25 hot dogs, one at a time, out of his bedroom window on the second floor of his house. After watching 25 hot dogs hit the ground and lay there, he might say, "You know, I'm beginning to think hot dogs can't fly. But I have to be sure. I remember trying to fly a kite as a kid and how hard that was."

When the scientist remembers trying to fly kites as a kid, he might think that it was hard to fly, but he knew kites could indeed fly because he saw his friends fly kites.

That gives him more ideas to try:

- He might wait for a really windy day and try throwing his hot dogs out the same window to see if the wind helps the hot dogs fly.

Super Science

- Or maybe he might try going to the 18th floor of a hotel and throw hot dogs out of that window because his next guess is that the hot dogs need to fall more distance before they fly.

IMPORTANT SAFETY TIP: Don't throw hot dogs out of a hotel window. Remember, you already know that hot dogs don't fly, and all that would happen is you would hear a lot of angry people on the ground yelling, "Hey, why is it raining hot dogs?"

So in conclusion, The Scientific Method is a great way to answer questions you have about the world around you and knowing and understanding how to use this process in the right way is an important first step in becoming a good scientist.

And now you know all about flying hot dogs…actually, you already know they can't fly…otherwise they wouldn't be waiting for that bus.

But what KIND of science is the Flying Hot Dog Experiment? That's coming up in the next chapter!

Cool Science Questions:

1. What are the steps of The Scientific Method?
2. What is a "hypothesis?"

Cris Johnson

Chapter 3:
The Four Sciences

In the last chapter, we spent a lot of time talking about whether or not hot dogs can fly. I know you thought it was silly. You know that I know you thought it was silly. And I knew that you knew that I knew...okay, I'm getting very confused!

I used the example of whether a hot dog could fly as an easy way to explain the Scientific Method because it's so important. Don't worry, we will be getting to the superhero stuff really soon, but covering some of this other stuff first is going to make learning about all of those cool experiments later even easier!

So, back to that flying hot dog...what kind of science is it?

To answer that question, we have to look at the four major kinds of science. Here they are:

Super Science

Life Sciences

Physical Sciences

Math Sciences

Social Sciences

So, here's a little bit of information about each one:

Life Sciences - This is the study of things that are alive. Pretty easy right? So, answer this question: if we study pet dogs, are we studying a life science?

If you said, "Well, yeah!" then you are correct! Why? Because dogs are alive! Here's the thing: When I do science assembly programs in schools like yours and I ask that question, kids sometimes don't want to answer. I think it's because the question of 'Is studying pet dogs a life science?' seems to be too easy of a question so sometimes we think, 'Well, that can't be the answer! That's way too easy!'

But that's the good news - sometimes an answer that seems too easy really is the correct answer!

So, right now, wherever you are, look around you and see if you can see living critters around you. (It doesn't matter where you are. You could be reading this in a bookstore, in your house, in school, on the moon, wherever.)

Cris Johnson

Maybe you see your own dog. If you were to study dogs, we already know that studying dogs can be considered a Life Science.

The name of one kind of scientist, a doctor, who studies dogs (among other animals) is a veterinarian. If my dog Stanlee is sick, I take him to visit my veterinarian to help him get better.

Or maybe there's some birds flying around your yard. They are alive too, so yeah, studying birds would also be a Life Science. A scientist who studies birds is called an ornithologist - try saying that three times fast!

What if you have a baby brother or sister? Yup, studying younger kids is a Life Science, too. I know, sometimes they might bug you, but your baby brother or sister is NOT a pest or space alien brought to Earth just to drive you bonkers. Believe me, I know...I have a brother eight years younger than me!

The study of babies is called...baby-ology. No, that's not right!

But a doctor who specializes in treating young kids is called a pediatrician. That's because as kids grow, they have different stuff going on inside their bodies as they develop and get older.

When you become a grown-up, you see a different kind of doctor: an old-a-trician! Hmmm, I don't think that's quite right.

Well, anyway, that's a little about Life Sciences.

Physical Science - Guess what? Studying anything that is not alive...is a Physical Science!

Let's look at some examples. Again, look around you, wherever you are. What do you see?

Maybe you are reading this book at night, and you can see these words because you've got a lamp or overhead light on. What powers an overhead light or desk lamp? Electricity!

Is electricity alive? Nope! So studying electricity would be a Physical Science. And the study of electricity is called electrostatics. That name makes a lot of sense!

What else is around you? Maybe you are reading this book outside while sitting in your backyard. Is the wind blowing?

Guess what - scientists study wind all the time! One way is through the science of meteorology, which is the study of how our atmosphere (the gases and stuff that surround us) affects our weather patterns.

So, when your mom or dad watches the news, the man or woman who tells them what the weather is going to be like over the next few days is called a meteorologist.

As you can guess, with both Physical as well as Life Sciences, there are a whole bunch of detailed sciences under those headings. There's just too much stuff to study for a person to say, "I'm going to study all of the Life Sciences!"

Nope, wayyyyy too much stuff to study!

Now, we have two more 'main' kinds of science to talk about.

Math Science - Yes, math is indeed a science!

When I do one of my science assemblies in schools, some kids are surprised to hear that studying math means you're a scientist.

And just like the Life and Physical Sciences, there are all kinds of math sciences! You have already heard of things like geometry and algebra, but math goes so much deeper than just the stuff you learn in math class. That's what we call the 'tip of the iceberg,' because when you look at an iceberg floating in the water, there's a MUCH bigger part of that iceberg that you don't see underneath the water.

To start with, there's science for studying the universe that is called Mathematics. When I say 'study the universe,' that means...*everything*. Air, water, space, frogs, newspapers,

Super Science

video games, other planets, oceans, your dad's smelly feet, I mean EVERYTHING.

Think about it. How many feet does your dad have? How many toes? There's mathematics right there. But it goes deeper, because whatever it is that is causing your dad's feet to smell can be measured and counted. When you smell something, that means there are little bits of whatever you are smelling that is going up your nose.

It's true! Smell a tasty burger from a restaurant? Little tiny bits of burger going right up your nose. If you picked some nice flowers for your grandma and she says, "These flowers smell so nice!" Guess what? Teeny-weeny pieces of flower went up her nose.

Soooo, that means if your dad's feet are really smelly, whatever it is that is causing his feet to smell, little bits of it are going up your nose.

Yuck!

What does this have to do with math, you ask? Well, if your dad's feet smell REALLY bad on a Tuesday, but not so bad on Saturday, that means there are fewer bits of whatever-that-smell-is on your dad's feet on Saturday.

How do you figure out how many fewer bits of smelly-feet things there are on Saturday as opposed to Tuesday?

Math.

Yup, we are talking about literally counting up the bits of whatever-that-smell is.

That's just one example of finding math everywhere to help us learn about the universe!

By the way, those teeny-weeny bits of things in the air are called "molecules" (or chemicals, depending which particles we are talking about). So there's another word you can use to show your parents what you learned in school or from this book.

Let's take another look at the situation with your dad's smelly feet. Imagine your mom saying, "What is making your feet smell so bad?"

Before your dad says anything, you helpfully say, "Little particles of stinky stuff called molecules. So we can call them Stinky Molecules."

Then there's statistics: that's the study of numbers to analyze data. In other words, scientists look at a bunch of numbers...maybe all of the people who watch the NFL. Then they look at those people and figure out how many boys watch the NFL, how many girls watch the NFL, how many grownups, and on and on. Then they take what they learned and use it to try to sell you stuff!

That's just one example. There's one more science to talk about, but before we get to it, I should address the elephant in the room.

No, I'm not trying to fool you: I know there really isn't an elephant in your room! Well, unless there is...then you have to send me a picture because that would be really cool!

"Addressing the elephant in the room" is a saying that means we have to talk about something really obvious and if we don't talk about it, many people would wonder why.

Here it is: Many of these sciences overlap each other. That means different kinds, or branches, of science might appear to be studying the same thing.

Think about the smelly feet example. You might think that is a Life Science thing because in our example, it's your dad's feet that smell and your dad is alive.

But as we just learned, it could be a Math Science study because mathematics studies everything in the universe and this makes sense because there are numbers and measurements EVERYWHERE, so if we are counting the Stinky Molecules on your dad's feet to try and figure out how many molecules it takes to make his feet stink so clearly it's a math thing...

To make things even more confusing, if whatever it is that's making your dad's feet stink is not alive, then studying those Stinky Molecules might be a Physical Science thing because they are not alive but they just happen to be stuck to your dad's feet...

Whew! My whole point in bringing all of that up is simply to show you that many sciences overlap and may appear to be studying the same thing.

Now that we have THAT out of the way, let's talk about the last science out of the Big Four.

Social Sciences - Social sciences are really cool because these sciences study how people act and behave.

That means studying these sciences helps you understand really important stuff! Here's just a short list of things social sciences can help you understand:

- Why you and your friends scream on roller coasters and other rides in theme parks
- Why just watching a movie that is filled with actors and fake special effects causes you to feel excited, sad, or happy
- What causes you to like certain foods but not others
- Why you might have trouble sleeping certain nights
- Why sometimes the best thing in the world is just snuggling up to your mom or dad underneath a blanket

That's just a super-duper tiny list of things Social Sciences help us understand. It's also my favorite science as I really like studying the mind and one thing I do when I'm not performing school assemblies in schools like yours or

Super Science

writing books like this is helping people with their problems.

Some people worry about things a lot and they have trouble stopping. We define it as feeling 'stressed out.' Other people are REALLY afraid of certain things - spiders, snakes, high places, and so on. If these fears are really bad, scientists call them 'phobias,' which means...the person is REALLY afraid of certain things.

Because I understand how peoples' minds work, I can help people figure out new ways of thinking about stuff so that they can feel better about themselves.

That's one of my jobs, called hypnosis, and it's one I enjoy a lot.

Another area of Social Sciences is studying stuff called optical illusions! In other words, we can do really cool things where we can trick our eyes so it seems like we are seeing things that could not really happen in the real world! Things like...seeing a hole in your hand...growing an extra finger...and making it look like one of your hands is passing right through the other!

By now, you might be thinking, "This is really awesome stuff and this Cris Johnson dude is such a good writer, but when are we going to get to the superhero stuff?"

Very soon! But let me first explain why I decided to talk about all of this other stuff before I got to the superhero powers.

This book is mostly about using science experiments to help us see what a superhero power would look like or at least seem like if people really had superhero powers.

To understand how to run these experiments, we have to understand the Scientific Method. Once we understand how to do experiments, it's really a good idea to know what kind of science we are doing, for a couple of really important reasons:

1. Understanding what science the superhero power we are experimenting with is involved with helps us understand everything much easier.

Think of it like this: Since you are reading this book, you know what the alphabet is. And when you learned the alphabet, you also learned what "vowels" are and what "consonants" are.

Let's pretend your teacher helped you learn the alphabet but did not bother to teach you about vowels and consonants. But then your teacher asked you to go through the alphabet, letter by letter, and count up all the vowels.

Since your teacher never taught you the difference between vowels and consonants, you would have a really tough time trying to do it.

Super Science

So all of this stuff in this chapter and the last one helps us understand all of this science stuff a lot better.

2. Understanding all of this can help keep you safe!

I will keep on saying this all through the book: There is nothing in this book or any other science book that is more important than you staying safe! Always remember that.

Now, in the rest of the book, we are going to imitate superhero powers like:

- Controlling the flow of liquid!
- Using the power of your mind to launch physical objects into the air!
- Create gooey matter using chemicals!
- And many more!

In fact, remember the optical illusions I mentioned: growing a third finger, x-ray vision, and passing a solid object through another?

We are going to do one of these experiments in the very next chapter!

Cool Science Questions:

1. What are the Four Sciences?
2. What is a pediatrician?
3. What does the field of electrostatics study?

Chapter 4:
X-Ray Vision

Yes! Now we are starting our very first superhero power in this book! Let's start with the first question from that list of questions that I said I would use for each chapter:

What is the superhero power?

This superpower is very cool - seeing right through solid objects! Many superheroes have this power and you can imagine how useful it would be to see through things: maybe the superhero can use this power to see where the bad guys are hiding. Or maybe somebody stole something, and the superhero uses his powers to find out where the item is being hidden.

Materials needed:

For this experiment, all you need is your two hands, your two eyes, and either an empty cardboard toilet paper roll or a regular sheet of paper that you can roll up into a tube.

Super Science

What is the experiment?

This experiment is really easy! You are going to make it seem like you can use x-ray vision and see right through your hand! In fact, when this experiment is talked about, it's often called the "Hole in Your Hand" experiment.

Here's how you do it: Sit or stand in one location. In other words, you can't do this experiment while you are running around!

While you are sitting or standing, look at something on a wall in front of you. You want to look at something kind of far away, so you don't want to be so close to the wall that you could stick out your tongue and lick the wall. (ewwww, gross!) And you don't want to be close enough to the wall where you could hold up your arm and touch the wall. You want to be at one end of the room and look at the wall across from you on the other side of the room.

Then you pick up that toilet paper tube and cover one eye with it. While you do this, keep both eyes open! Be careful not to poke yourself in the eye with the toilet paper tube.

Next, cover your other eye with your other hand. Keep both eyes open!

If you have done this correctly, you should now be seeing what looks like a hole right through your hand!

I have noticed that when I do this, the hole looks like it's just on the edge of my hand. I have found that if I move the hand that is covering my eye a couple of inches away from my eye, that helps it look like the hole is more in the center of my hand. It's so neat!

This is considered an 'optical illusion' because your brain is receiving information and trying to convert into an image for you to see that doesn't make any sense in reality. In other words, we are tricking the brain into causing us to see something that is not really true.

Why is it a "superhero power?"

As you might guess, we really CAN'T see through things. It's just not the way our eyes work. Because it's not something people can do naturally, if there really was a superhero who could see through solid objects, that would be considered extraordinary! As in "extra-ordinary," like "an extra ability above the ordinary abilities we have."

However, using technology, scientists have developed ways to see through solid objects. If you go to see your dentist and he or she takes x-ray photos of your teeth, the doctor can then look at your photos to see if your teeth have any cavities.

When you go to the airport to get on a plane, you go through security and the people there take x-rays of your body to check and make sure you are not bringing anything unsafe onto the plane. Same thing when they scan your bags - they are able to see into the bag to make sure nothing dangerous is being put on the plane.

Then there's something called thermal imaging: That's using the heat that something gives off to locate it or take a picture of it. Police might use something like this if they are trying to find bad guys in a building: Using the thermal imaging cameras will allow them to track the bad guys using the heat their bodies give off and they can see these images right through walls!

So in that sense, scientists have looked at what superheroes can do in stories and figured out ways to use technology to bring that power to life!

Where should I do this?

This is an experiment you can do almost anywhere! As long as you can safely sit or stand in one place, you can do this experiment.

Safety:

I already mentioned one thing to be careful of in this experiment: being careful when putting the toilet paper tube or rolled up piece of paper over your eye. Your eyes are very delicate - that means that it is really easy to damage them - so anytime you put anything anywhere near your eyes, you want to be super careful!

Also, earlier I wrote that you can't do this experiment while walking or running. That's because while you are walking or running, there is no way you would be able to hold your hand steady enough to keep from jiggling your hand and pushing that tube into your eye.

On top of that, if you were holding a tube over one eye and you were covering your other eye with your hand, you could easily trip and fall if you were walking around while doing this.

Remember, safety first!

What are some others ways to experiment with this:

You can change the experiment by remembering the Scientific Method: Ask a question!

What are some questions you could ask to change this experiment? Here are a few I thought of as I was writing this chapter:

- *What would happen if I used a longer tube?* Maybe instead of a toilet paper tube, you could try a paper towel tube, which is a lot longer. What do you think will happen? Make a guess and try that!
- *What about doing the experiment with a toilet paper tube that still has all of the toilet paper still on it?* What would happen then? Instead of having a very thin tube over your eye, now you'd be trying it with a great big fat tube with all of the toilet paper on it.

Super Science

Make a guess - how would that change what you see when you do the experiment? Then do the experiment and see if you were right with your guess!

- *What would happen if you try the experiment with transparent or partially transparent paper?* The classic way to do the experiment is with a toilet paper tube, which is a piece of cardboard rolled into a tube. We can't see through solid cardboard. What if you tried this experiment with a piece of paper? Use a piece of paper was so thin that light still sort of came through, like the kind of paper that goes into a printer. Would using a thin piece of paper change how you see that hole in your hand? Make a guess and try it!
- *While covering one eye with your hand and covering your other eye with the toilet paper tube, what would happen if you slowly moved both away from you?* Would you still see a hole in your hand or would the illusion stop working as soon as you moved your hand and the tube away from your face? Make a guess - what do you think will happen?

Those were just a few things I thought of, but you might come up with even more ways to test how this experiment could be different by changing some of the steps. That's what makes science so much fun!

The science of the experiment:

This experiment is neat because it creates this illusion by taking advantage of something called "binocular vision," which just means we have two eyeballs to see stuff. We could actually just have one eyeball in the center of our heads and get along just fine, but having two eyes allows us to have better depth perception. That means our two eyes help us judge how far away or how close something is, which makes it easier for us to do things, like catch a baseball as it flies toward us, or eat that French fry as our hand moves it closer to our mouth.

Binocular vision also gives us a better field of vision than just one eye. Field of vision means how much stuff we can see around us at any one time. To understand field of vision better, try this:

Look out your bedroom window, and keeping your head in one position, and without moving your eyes, notice how much stuff you can see. Then, cup your hands around your eyes so you can still see, but the sides of your vision is blocked by your hands. So, cupping your hands around your eyes limits how much you can see...in your field of vision.

So with binocular vision, your eyes are seeing slightly different things because they are spaced apart from each other. Each eye is taking in information that is different, but since the different information is just a few inches of space, it's not a BIG difference. Your brain takes the information that both eyes are receiving and combines that information so you "see" one thing.

By covering one eye and putting the tube over your other eye, you are causing each eye to now see drastically different things, so when your brain tries to combine the images you "see" something that is not true: the hole in your hand!

Similar experiments:

There are all kinds of fun optical illusions you can try either by looking at stuff online, or reading more science books...and even this book! In this very book, you'll be learning two more fun optical illusions: one makes it look like you are growing an extra finger and the third optical illusion in this book makes it look like one of your hands is magically passing through the other! Stay tuned because these are FUN!

Which of the 4 sciences does this fit:

This experiment could be considered part of the Life Sciences because we are talking about living beings (us!) doing something with our sight, so that's pretty easy to understand.

However, we might also be able to put at least some optical illusions into the Social Sciences category because optical illusions can sometimes be considered part of psychology, which is also part of Social Sciences.

Psychology, by the way, is the study of how people think and behave. For instance, if there was a joke that you thought was really funny and you decided to tell it to people and after they laughed, you asked those people questions to find out why they thought that joke was funny, that means you are involved in psychology - trying to understand why those people laughed.

The website where you can see the experiments:

www.SuperheroScienceBook.com

You will need your parents to fill out the contact form to get access to the videos. Make sure you only go on the Internet with your parent's permission!

For our next experiment, we are going to look at a superhero power where it looks like we can create matter out of thin air!

Cool Science Questions:

1. What is psychology?
2. What is thermal imaging?
3. What is binocular vision?

Super Science

Chapter 5: Creating Solid Matter Out of Thin Air!

This experiment is SO fun because it's gross! Well, that's not the only reason it's fun. It actually looks quite magical as it happens, like there's no way possible that what you are seeing could actually exist!

Let's get to it!

What is the superhero power?

There have been many superheroes who have the ability to create matter out of thin air! Oh, this is important: matter is everything around you. The shoes you wear? Composed of matter. YOU are composed of matter! That mac and cheese lunch you love? Matter. That hot dog that you hopefully did not throw out the window? Made of matter.

Some superheroes have the ability to create ice out of thin air. In that case, 'creating ice out of thin air' means just that: The superhero has the power to freeze the air, changing the air from a gaseous state to a solid state! (More on states of matter later in this chapter)

Other superheroes have the ability to duplicate themselves to have more ways to fight a bad guy or army of bad guys, so in a sense, they are creating flesh and blood with their power.

Mad scientists often have the intelligence and technology to create solid matter in various forms. In that case, it's not so much a super power but more that just they are really smart, and have invented machines or technology that does the 'super power' for them.

The point is, in comics, the ability to create matter is a popular one that has been around a long time!

In this case, what we are going to do is create a huge blob of foam by mixing together some chemicals.

This is called Elephant Toothpaste, because the material created looks like a huge amount of toothpaste, big enough for an elephant to use. And when it's done correctly, it just pours and pours out of a bottle, flowing UP and out of the bottle, so after taking a cup or two of ingredients, we are creating liters and liters of material, so it looks like we are creating this matter out of thin air!

Materials needed:

You may have seen versions of this on YouTube or science shows on TV, and the scientist on those shows usually create so much elephant toothpaste that it completely covers a table or in some crazy cases, fills up part of a room!

That does look really cool, but it uses versions of some of the chemicals that can be really harmful to you. This version is much safer and when you try it, you'll see that while it doesn't fill a room, it's still really cool! And remember at the end of this chapter, you'll see the webpage address to see a video of me doing this too.

Here's what you'll need:

- 2 Tablespoons warm water
- 1 Teaspoon yeast
- Food coloring
- Hydrogen peroxide – Either 3% or 6%
- Dish soap
- Empty 16oz clear soda or water bottle
- Funnel
- Eye goggles or safety glasses

Here's some things to think about with these ingredients...

First, you can get everything you need from the grocery store, so this is super-easy to get set up!

The first time I did this experiment, I used water that was way too warm. I actually heated it up by using the microwave! As it turns out, that's far too warm for this experiment because the yeast are these little one-celled critters that need food and warm water to survive but if the water is too hot...uh...the yeast, these itty bitty living creatures, don't like that.

So when I poured my yeast into the bottle, I did not get elephant toothpaste. I just got a container of sludge that just sat there. Yuck!

That brings me to the yeast...this is what bakers use in cakes and breads which cause it to rise, that is, it gets big and fluffy. Without yeast, baked bread would just be doughy and gross. Yeast is awesome, but it IS alive, so whenever your mom or dad bakes, they are putting live things into your food. That is so weird!

For the dish soap, I found that Dawn dish soap works the best but you can use any brand you can get your hands on.

For the bottle, every scientist I have ever seen says to use a 16oz soda or water bottle. But then every video I saw showed the same scientists using a 2 liter bottle, which is obviously much bigger.

The hydrogen peroxide is stuff that your parents probably have in the medicine cabinet or somewhere that they use in case you get a skinned knee or a cut. They pour a little on where you hurt yourself, and boy does it ever sting! But that stuff keeps you from getting an infection. That means all kinds of germs could get into your cut and then the cut would get worse and look all puffy and gross. So the peroxide helps.

Super Science

I wound up using the bigger two-liter bottle as well and then I just doubled the amount of the ingredients. If this is your first time doing this, I recommend doing it just like I'm writing it, with the sizes and ingredients listed above. You can always try different amounts of ingredients and container sizes later.

You definitely want to use a funnel to pour the yeast and water into the bottle because the funnel makes it easier to get it into the bottle without spilling...plus you get more of it into the bottle and that means more of a reaction, which means more toothpaste!

What is the experiment?

Now it's time to do the experiment! To prepare, here's what you do:

1. Using the funnel, pour ½ of a cup of the peroxide into the bottle.
2. Squirt some food coloring into the bottle. (Later I'll give you some tips to really make your toothpaste pretty, but to keep these instructions simple...just squirt some into the bottle.)
3. Squirt a bunch of dish soap into the bottom of the bottle. You want a tablespoon or so if you want to be super-precise, but just giving the bottle a squeeze will put the right amount into the bottle.
4. Next, pour the two tablespoons of water into a small cup, and then add the yeast. Stir the yeast into the water until it is completely dissolved.

"Completely dissolved" means there are no more lumps of yeast and the water / yeast mixture is smooth, so it looks like muddy brown water. Keep stirring for another minute or so. This gets the yeast up and ready to rock!
5. Put your safety glasses on and place the bottle on a surface where it's not going to fall over.
6. Using the funnel, pour the yeast / water mixture into the bottle and wait: You should see a column of toothpaste-looking goop rise up out of the bottle and all over the table surface where you are doing the experiment.

Voila! You've just created a goopy mess, that looks like it came from nowhere! Oh, and did you know the word "Voila" means "there it is!" I just learned that!

Why is it a "superhero power?"

This can be considered a "superhero power" because humans (you, me, and any other people reading this book. If your dog is reading this book...I'm scared) do not have the ability to just create some kind of matter out of thin air.

Now that we have established that, we DO know that using chemistry and other sciences, we can create matter out of other material, so this is a superhero (or super villain) power that can in some form seem to exist in the real world. But of course, everything is exaggerated and made bigger and crazier in comics.

Super Science

Where should I do this?

Because this experiment makes such an awesomely big mess, this is definitely something you want to do outside! You also want to be sure you do this on a surface like a driveway or tabletop that is stable and not likely to allow the bottle to tip over.

Second, you want to make sure that wherever you do this, there is nothing that is going to get stained and get you in trouble!

Finally, you'll need to remember that because this does create a big mess, you'll want to do this where it's easy to actually clean it up. This is not an experiment I can ever do in a school assembly because it creates such a BIG mess.

I do a lot of my school assemblies in places like gyms, and I do not want to mess up the floor for the gym teacher!

Safety:

This experiment is really fun, but like anything, if you are not careful, bad stuff could happen, so here are a few things to remember.

The hydrogen peroxide stuff I recommend using the first time you do the experiment is the 3% solution in the bottle I wrote about earlier. In the video, I used a 6% solution: It was twice as intense and produced a bigger reaction...more goop!

But because the 6% solution is twice as intense, it can cause skin irritation, so please use the 3% stuff first. Besides, the 6% is harder to find. Hair dressers use it all the time in hair salons. In fact, that's how I got mine: my wife owns a hair salon!

This experiment is actually pretty safe, especially if you use the 3% solution, but it's always a good idea to wear safety googles or glasses whenever you are using chemicals because any of the chemicals such as peroxide, dish soap, or even food coloring could hurt your eyes.

Speaking of food coloring, if you use it straight out of one of those little 'squeezy' bottles it comes in, it's pretty concentrated. So if you spill any on your clothes or your mom or dad's favorite table cloth or rug, they are going to be pretty mad about it because it would leave a stain that's really hard to get out,

Same thing with your skin: if you get food coloring on your skin, you'll be stained that color for quite a while. Eventually it will go away but you would have to scrub and scrub over and over, and if you're anything like me, you don't take a bath more than once or twice a week.

I'm just kidding! It's silly to think that anyone would need a bath THAT often.

But if you spill green food coloring on yourself, you'll look like you have goblin skin for quite a while!

Super Science

By the way, "concentrated" means there are a whole lotta molecules packed together in a tight space by themselves. If you were to squirt just two drops of food coloring into a big bucket of water, it would no longer be very concentrated at all. In fact we could say that because there were only two drops of food coloring in that big bucket of water, the food coloring was now "diluted," meaning that a little bit was spread out far.

The other chemical used in this experiment, dish soap, is also something you don't want to get in your eyes…or your mouth! Gross!

Finally, the chemical reaction does generate some heat. It's called an "exothermic reaction," which is just a fancy way of saying that it's…uh…a chemical reaction that generates heat. (Bonus point for you if you look up the word "redundant!")

Touching the bottle afterwards to feel that heat is really neat, but again, safety first so please be careful when handling any of the material before, during, and after the experiment.

What are some others ways to experiment with this:

If you want to try to make your elephant toothpaste look extra pretty, you can try this!

When I do the experiment, I pick two colors that look good together, like purple and blue. Then I squirt some of one color into the bottle, but I try to aim it so the food coloring that gets squirted into the bottom sort of lands along the left side of the bottom of the bottle. Then I do the same thing again but try to squirt it in so the food coloring lands in the right side of the bottom of the bottle.

Then I take the second color and do the same thing, but squirt it so it lands at opposite ends of the bottle.

Think of it like this: let's pretend your mom or dad has cooked you dinner and they made chicken, noodles, apple sauce, and chopped up hot dogs. (I know...that dinner sounds AWESOME!)

Anyway, your mom probably won't mush all of that food together. She will probably put the chicken on one section of your plate, then the noodles in another section, then the apple sauce, and finally the chopped up hot dogs. She will probably put each food on your plate in it's own "quadrant."

A "quadrant" is one quarter of a circle. Each quarter is divided by a real or imaginary line. In the case of our dinner plate, there could be little raised dividers in the plate to keep your apple sauce away from your hot dogs, or if it's just a regular plate, your mom or dad probably just did their best to divide those four foods.

So what you are doing is trying the squirt two colors of food coloring in a way that they don't mix with each other.

It will make it so your toothpaste comes out with really cool looking stripes!

I've tried it with three different colors, but they wound up mixing together too much and my toothpaste just looked all gray and gross.

You can also try increasing the amount of the yeast, Peroxide, and dish soap. You can see how changing these amounts of the ingredients will change how much or how fast the toothpaste will come out.

I've also cut off the top of the bottle with a pair of scissors, so when the toothpaste comes out, it's coming out of a bigger hole so you get a much thicker column of goop!

Of course, always be careful with scissors!

The science of the experiment:

In school you've probably already learned about states of matter, but just so this book leaves out nothing, here goes: Everything around you is made of matter, like I mentioned a couple of paragraphs earlier. We can define matter by it's state.

For instance, water is a liquid state of matter: liquids take on the shape of whatever they are in or on. If you have water in a bucket, it takes on the shape of the bucket. If you throw it on the floor, it's going to spread out and get flat like the shape of the floor.

Solids, on the other hand, retain their shape. If you throw a big rock into that same bucket, it's going to still be the same shape even as it sits in the bucket. If you throw that rock on the floor, it's going to still be the same shape.

A gas fills whatever container it is in completely. If you release some air into a room without air, it's going to spread out and fill that room. If you put that air into that bucket, it's going to fill that bucket but it's also going to spill out into the area around the bucket too because without an airtight lid, there is nothing to hold the air inside the bucket.

We used to believe there were just three states of matter, but the cool thing about science is that we are always trying to get better and be more accurate, so things we believed years ago have changed.

The fourth natural state of matter is plasma. Plasma is a gas that also has a LOT of energy. The atoms (the smallest measurement of matter) in gas are already free to move and jump around. Well, when energy is added to the atoms in gas, those same atoms get *really* excited and bounce around super-fast!

Among other things, scientists use plasma to make fluorescent light tubes!

The fifth natural state of matter is Bose-Einstein condensates. It's actually what happens when a dilute gas is made to be incredibly cold, near absolute zero.

Super Science

A "dilute gas" is a gas that has been diluted, kind of like what happened to the ketchup bottle when I was a kid. Once, my mom made French fries for dinner and we realized we only had a little bit of ketchup for the fries. So my mom poured a little bit of water into the ketchup bottle. The ketchup wasn't as thick as it should have been, but it gave us more to use for dinner. A diluted gas is sort of like the watery ketchup, but it's a lot harder to do with a gas.

"Absolute zero" is the coldest temperature believed to be possible. The temperature is –273.15° on the Celsius scale or –459.67° on the Fahrenheit scale. That is COLD...like wearing two pairs of socks cold, at least.

There are other states of matter, but let's get back to that toothpaste.

What causes the toothpaste to rise up like that?

It's a few different things that work together.

First, what is the peroxide made of? It's a combination of hydrogen and oxygen plus some liquid. This experiment separates the hydrogen molecules from the oxygen molecules. The yeast is what causes this to happen very quickly - in fact when a substance causes a chemical reaction to happen quickly, it's called a "catalyst."

When the oxygen (a gas) separates from the hydrogen (another, lighter, gas), the oxygen always wants to rise and escape the liquid.

Think about when you look at a glass of Pepsi or Coke: you can see hundreds of little tiny bubbles rising to the top and popping. That's the trapped carbon dioxide gas looking to escape.

With this experiment, it's a similar idea: The yeast helps separate the oxygen, which is a gas, and cause the bubbles to rise super-fast! But just causing the bubbles to rise rapidly doesn't give us our toothpaste goop.

That's where the dish soap comes in! The soapy liquid adds flexibility and strength so the bubbles don't pop as quickly. Think of the dish water your mom or dad or whoever uses to clean the dishes at your house. The bubbles in the soapy water last several minutes as the dishes are being scrubbed.

When you think about it, elephant toothpaste is a big column of bubbles!

Similar experiments:

There are similar experiments you can explore in other science books, too. For instance, you can build an Alka-Seltzer Lava Lamp or make a Lemon Volcano.

However, later on in THIS book, you'll learn how to make Bubble Snakes which are REALLY neat, and you'll learn my personal favorite science experiment called Diet Coke & Mentos, so there's a lot of really cool stuff to look forward to!

Which of the 4 sciences does this fit:

This experiment is definitely in the Physical Sciences category. Like I mentioned in the chapter on the Four Sciences, each one of these main categories of science each have hundreds of other specialty science in them.

In this case, this experiment falls into the area of Chemistry, which can be defined as "the composition and properties of substances and of the changes they undergo." That's a fancy way of saying that Chemistry is the science of studying how stuff is made up and how it changes when things happen to it.

In other words, it's fun!

The website where you can see the experiments:

www.SuperheroScienceBook.com

You will need your parents to fill out the contact form to get access to the videos.

Make sure you only go on the Internet with your parent's permission!

Cris Johnson

Now, onto a fun experiment where it looks like you can stop a falling object by nothing more then the power of your mind!

Cool Science Questions:

1. What is yeast used for in everyday life?
2. What are three states of matter?
3. What is a "catalyst?"

Super Science

Chapter 6:
Telekinesis:
Stopping a Falling Object

This experiment really surprised me the first time I saw it! A magician did it with some kids on stage and I was sure the performer was using some fancy hidden magic gizmo, but it was nothing but science!

What is the superhero power?

As the title of this chapter says, this experiment looks as though you are using magic or some crazy mind power to stop an object from crashing to the ground!

Here's how it looks: You take a water balloon that is tied to a string, with a small metal ring at the other end, and drape the balloon over a pencil or wooden spoon.

You let go...and before the balloon crashes the ground, the metal ring has wrapped itself around the wooden spoon and stopped the balloon in its tracks!

Cris Johnson

Materials needed:

The materials for this experiment are super-simple (there, I did it again):

- A water balloon filled with water
- A length of super-strong string about 3 feet or so long - you want super-strong string, like kite flying string
- A small metal ring - either a keychain ring or a shower curtain ring works well
- An unsharpened pencil or a wooden spoon with a rounded handle

Oh, and in case you have never filled a water balloon before, here's how you do it: take the open end of the balloon and wrap it around the opening of the kitchen faucet. If you let go of the balloon and it dangles from the faucet, you know you have done it right.

Then you sl-o-o-w-w-ly turn on the water and fill up the balloon. When the balloon is filled, quickly turn off the water. You have to pinch the opening of the balloon very tightly as you take it off the faucet to keep any water from leaking out and spraying everywhere.

By the way, I didn't go to a lot of outdoor parties as a kid, so if you have never filled a water balloon before, it's okay. I didn't fill my first water balloon until I was an adult in my 40's!

Super Science

If you want to be extra careful and prevent possible spilling, you can take two balloons and shove one inside the other, and then fill up the inner balloon with water. That will give extra strength to the water balloon.

After you fill the water balloon, tie the end with the opening into a tight knot.

Then take the water-filled balloon and tie one end of the string to it.

Tie the other end of the string to that metal ring.

That's it - now you're ready!

What is the experiment?

This experiment is super-simple: Stand up straight and with one hand, hold the pencil parallel to the ground.

"Parallel" means that two or more things lie or move in the same direction and remain the same distance from each other no matter how much you move.

Think of parallel like this: let's pretend a bird is flying above you and you run along below the bird, if you run in the same direction as the bird flies, you are parallel to each other.

So, hold the pencil parallel to the ground. Then, take the balloon-and-string contraption you made and lay the balloon over the pencil so the weight of the balloon is dangling toward the ground, but if you hold the string by the ring, the balloon stays in place.

When you are ready, let go of the ring! If the experiment works, the balloon should fall to the ground...but be stopped from hitting the ground by the ring, which has wrapped itself and some of the string around the pencil!

Why is it a "superhero power?"

People don't have the power to do anything with their minds and thoughts other than think about stuff.

That means, no matter how hard you think "Don't fall," that balloon is going to hit the ground...unless something stops it.

In comic books, things are different. There are all kinds of superheroes who have what's called the power of "Telekinesis," which is the ability to move physical things just by thought.

It's an amazing power and superheroes use it to move objects away from people to keep them safe or use the power to 'throw' things at the bad guys by using their mind in battle.

Telekinesis is something that some scientists believe might be possible and they do test these abilities that some people claim to have in laboratories.

Super Science

So far, no one has ever been proven to be a real superhero with the ability to move things with the power of their mind…

But one guy got close.

Many years ago, a magician by the name of Steve Shaw went to some scientists and claimed he had the power to cause metal objects to bend and twist with the power of his mind.

Some scientists were so impressed by Steve that they told the public, "We have found a person who can really do these amazing things by the power of his mind!"

Really, Steve was using very clever magic tricks to fool the scientists. His methods were really clever, and years later, I learned some of his methods and some people (not scientists) thought I might have those powers too!

Steve eventually revealed to the scientists that he was just using magic tricks to fool them. His friend, James Randi, had convinced Steve to try and fool the scientists.

Steve didn't do this to embarrass the scientists. He did it to remind the scientists to always use the Scientific Method to make sure that what they were testing was being done properly…because some of the scientists wanted so much to find people who could do these amazing things for real that they sometimes forgot to make sure they were testing everything properly and getting true answers from their experiments and not just what they wanted to see!

Where should I do this?

To be totally safe, you should definitely do this experiment outside! Breaking a water balloon really isn't dangerous, but if you get water all over things in your house, your mom and dad might get really mad.

I have done it indoors and sometimes when the balloon hits the floor, I'm very surprised that it doesn't break.

If you decide to do it outside, doing the experiment over grass can be very helpful because if the balloon does hit the ground, the soft grass can help prevent the balloon from breaking.

If you do the experiment on a sidewalk or driveway, there's more of a chance of the balloon breaking if it does hit the ground.

I will write more about this in the "Safety" section.

Safety:

In the previous section, I mentioned a few times what could happen if the balloon does hit the ground. That is because when I have done this experiment, it has not worked every single time. I'll explain what I think is happening in the "Background" section, but keeping things safe is a good idea even if it worked every time.

Super Science

Another thing to remember...in my earlier instructions, I mentioned using a three-foot piece of string. That will work if you are tall enough. Let me explain...

You know already that people come in all kinds of different shapes, sizes, and heights.

So I'm a little more than five and a half feet tall. That means if I hold my arm straight out in front of me and I hold that metal ring by the end and let the balloon dangle on a three foot piece of string, the balloon will not touch the ground unless I let go of the string.

Before you do the experiment, try holding the ring in your hand, with your arm straight out in front of you. If the balloon does not touch the ground, your string is the right length. If the balloon does touch the ground, it just means you'll need to use a shorter string.

The first time I saw this experiment, the magician was doing it by standing on a chair so the balloon was up high! It looked really unsafe, so I would never stand on a chair to do this.

Also, if you want to really take precautions and keep the balloon from bursting, you can just fill the balloon part way with water.

The more you fill the balloon, the more the balloon itself has to stretch to hold all that water. So if you fill the balloon up completely with water so the balloon is so stretched out that it can't hold anymore, the wall of the balloon is under a lot of tension and it won't take much for it to pop.

By filling the balloon up only half way, you're taking a lot of pressure off of the wall of the balloon.

What are some others ways to experiment with this:

You can experiment with this in a lot of ways! Remember when I mentioned that this experiment did not work every time for me? I think it might be because of the length of string I use. I still have to do some experiments with different lengths of string to see if that makes a difference.

You can try different lengths of string to see if that makes the experiment work more often or not.

You can try swinging the balloon back and forth before you let go to see if that helps the experiment or not!

Maybe you could try filling the balloon with things other than water, like jelly! Of course, then you REALLY want to be careful because no one wants to be hit with a Jelly Balloon that splatters! Gross and sticky!

The science of the experiment:

The major scientific principle that makes this experiment work is gravity. (There's another principle at work but we will get to that in a moment.) Gravity is the force that holds you to the earth. Because the Earth and it's moon are similar in terms of the kind of matter, but very different in size, the Earth's gravity is stronger than the moon's gravity. For instance, if you went to the moon, which is roughly one quarter the size of the Earth (where we live), you would weigh about one sixth of what you do on Earth! (I've been meaning to lose some weight...maybe I should live on the moon!)

If you throw an object into the air, it falls back to the Earth, gravity is what's causing it to fall down.

Where it gets really interesting is this: if you hold a bowling ball and a feather in the air at the same height and let go of both at the same time, and the bowling ball hits the ground first, you might think that gravity is causing the heavier object to fall faster than the lighter object...right?

Nope. Gravity pulls both objects back to Earth at the same rate. The reason why the feather reaches the ground more slowly is because of air resistance: The shape of the feather is affected by air resistance, causing it to fall more slowly because the feather's shape is sort of like a sail on a sailboat. On the other hand, since the bowling ball has more mass, there is much less air resistance.

So, what causes the ring to wrap around the pencil? A couple of things: "pendulum," and "friction."

Because you hold the ring end of the string stretched out parallel to the ground, the ring begins to fall when you let go. When the ring begins falling, it's not falling straight down but rather at an angle.

The ring becomes a "pendulum," which is a body of mass supported by a string at one point, swinging at an arc.

Because the balloon falls at such a fast speed, the length of string on the side of the pencil with the ring decreases really fast.

As the length of string decreases as the pendulum swings, it's swinging speed increases, causing the ring to swing all the way around in a circle, multiple times, going faster and faster.

To use a math term, the **radius** of the circle gets smaller and smaller as that ring is wrapped around the pencil.

As this process continues with the string getting shorter and shorter, the speed of the swing gets faster and faster, until the ring is wrapped tightly around the pencil.

By the way, "radius" is the measurement in a straight line from the center of a circle to the outside edge of the circle.

That brings us to the second principle at work here, which is friction. "Friction" is the resistance of motion when two objects rub up against other. The less smooth or slippery two objects are, the more friction will be present.

What this means is that the string is wrapped tightly around the pencil (and itself) so the bumpy uneven texture of the string holds it in place and keeps it from slipping free.

What looks like an amazing superhero power is actually two simple pieces of science!

Similar experiments:

There are all kinds of other fun experiments you can do with balloons!

Here are just a few:

- Try this same experiment with different objects! The weight ratio needs to be about 14:1. That means the heavy object needs to be about fourteen times the weight of the light object. I first saw this experiment done with a ring and water balloon, and as I researched it, I found out people were doing it with things like fifteen metal washers: One washer at one end of the string and fourteen washers string together at the other end of the string. As long as you get that weight ratio right, you could do this experiment with all kinds of objects!
- Blow Up a Balloon Without Blowing - This experiment is so much fun and helps you learn about acids and bases! As the name suggests, this is really weird as you literally blow up a balloon without using your breath!

- A Balloon-Powered Car - This experiment involves using balloons to power a car, and to make it even more challenging, use recycled materials!
- Float a Balloon-Powered Boat - This is very similar to the Balloon-Powered Car!

Your parents or teacher can help you look these up online, or in books in your school's library. Also, I will soon be publishing an entire book on simple machine science!

Which of the 4 sciences does this fit:

This experiment is within the Physical Sciences category. Digging further, this experiment is part of the field of Gravitation, which is the study of gravity.

One of the most important people in this field was a guy by the name of Isaac Newton, who lived all the way back in the 1600's.

There's an old story saying that Newton discovered gravity when he saw an apple fall from a tree. He started wondering what caused the apple to fall, because no person or animal had touched the apple.

No one knows if this story is true, but regardless of what happened, he began working on ideas and experiments that led to our understanding of gravity!

Newton is considered one of the most important people in the history of science, and formed a lot of our key understandings of not only how stuff works on Earth but also the entire Universe!

One of the great things about studying the history of science is that it can tell us a lot of things about how we developed science that we use today...and we can learn from any mistakes we made, and try not to repeat them in the future.

This experiment also fits into Math Sciences because of the pendulum principle. A So two different branches of science in one cool experiment!

The website where you can see the experiments:

www.SuperheroScienceBook.com

You will need your parents to fill out the contact form to get access to the videos. Make sure you only go on the Internet with your parent's permission!

In our next chapter, we will be doing an experiment that will simulate a superhero power of causing objects to launch really high in the air!

Cool Science Questions:

1. What is Isaac Newton famous for discovering?
2. What does parallel mean?
3. What does radius mean?

Cris Johnson

Chapter 7:
Telekinesis:
Make Objects Fly into the Air

This is one of those experiments when I saw it for the first time, I thought I was seeing an incredible magic trick! This experiment is easy, fun, and looks amazing! Best of all, there's nothing 'messy' to worry about, although like any science experiment, there are ways for you to get hurt, so read this chapter all the way through before you try this!

What is the superhero power?

Once again, we are talking about telekinesis, or moving things with your mind, but in this case we are not talking about stopping a falling object...nope, this time, we are using telekinesis to cause objects to fly up into the air 20 feet or higher!

Materials needed:

For this experiment, you really only need a few things:

Super Science

- A basketball
- A tennis ball
- A safe place where there is a smooth driveway or pavement

What is the experiment?

Go to the driveway, bringing the basketball and tennis ball with you. Hold the basketball in both hands, with your arms extended in front of you. Let go of the basketball and watch as it bounces off the driveway. You'll see that it only bounced back up around half of the distance it fell.

In other words, if you let go of the basketball when it is three feet off the ground, it's only going to bounce back up between one and a half feet to two feet. Something is happening to take away some of the basketball's bounce!

Try the same thing with the tennis ball: if you bounce it onto the pavement or driveway, you'll see that it does not bounce all the way back up. What's going on?

I'll explain everything soon. First, let's do the crazy superhero experiment part!

For the superhero power, here's what you do: On the driveway or pavement, stack the tennis ball on top of the basketball. You'll have to hold the tennis ball in place because if you let go, it would roll off.

Hold both stacked balls in the air with your arms straight out in front of you. When you are sure the tennis ball is stacked straight on top of the basketball, let go!

Since the basketball is on the bottom, it will hit the driveway or pavement. Then watch as the basketball bounces back up a couple of feet, but the tennis ball flies up high in the air, 20 feet or more!

It looks just like you used some kind of weird superhero power to cause that tennis ball to fly up so much higher than anyone would have guessed!

Why is it a "superhero power?"

This looks like a superhero power for the same reason that the Balloon Drop looks like a super power: people do not have any ability to cause objects to stop falling or jump high into the air just by using their brains and thinking about it.

The cool thing is people have invented all kinds of machines using science to do the same thing.

For instance, if you have ever been to a professional sporting event, like the NBA, sometimes people come out during breaks or half time and launch t-shirts into the crowd using what's called a T-shirt Launcher or a T-shirt Cannon!

Super Science

If you have ever used a rubber band to launch a wadded-up piece of paper into the air, you too have used science to simulate a superhero power!

When you think about it, a CAR is a way to transport people from one place to another very quickly, because a car can easily go faster than the fastest running person who ever lived! Cars, trucks, buses, airplanes, and even your bicycle were invented to help people get from one place to another faster than before those things were ever invented.

Here's the deal: People use science to invent a lot of things in order to make our lives easier, safer, or healthier. Scientists invent medicines to help us get better from being sick. Scientists invent machines to do work for us faster and safer. Scientists even invent things to entertain us and help keep us from being bored!

I mention all of this because it might remind you of the very first sentence in this book: Remember when I asked if you thought science was boring?

Try this when you are done reading this book. Look at everything around you. Books, lamps, ovens, refrigerator, cell phone, iPad, TV, computer, your shoes, the locks on your home doors...can you see anything around you that was made by people that DIDN'T come from science?

Cris Johnson

My hypothesis is that you won't be able to find anything in your house that was not invented, designed, or built involving science!

When you start to think about all the amazing things we have in life because of science, you start to realize just how much FUN science can be!

Where should I do this?

This is definitely an experiment you want to do outside! Those balls can bounce pretty well off of your kitchen floor, but they will bounce too high and might damage something in your house, and then your parents might get real mad and say, "No more science in the house!"

This has happened to famous scientists before. In fact I heard that when Isaac Newton was young, after he discovered gravity, every time something fell down, his mom got mad at him for discovering something that caused so much trouble.

Okay, I made that part about Isaac Newton up...but this is an experiment that can really wreck a lot of things so you want to do this outside...away from anything that might break, like windows, car windshields, and cats!

Also, you definitely want to do this on a hard surface like pavement or a driveway. My guess (my hypothesis!) is that since grass is soft, the balls will not bounce nearly as much and you won't get to see that tennis ball bounce up very high at all.

Safety:

This is a pretty safe experiment, because there are no dangerous chemicals or gross things you can get into your mouth!

The 'safety' thing really comes down to where you do this. I've described the experiment being done on a driveway. Think quickly! What's often on a driveway? If you said "a car, truck, or van" you are correct!

I don't think bouncing a basketball on a car is going to hurt the car much, but it could scratch the paint or do something else to make your parents mad so always do this experiment far away from any vehicles.

Same thing with the windows of your house: I have found that unless the tennis ball is EXACTLY on top of the basketball, when it goes up, it often goes at an angle: off to the right or left or even behind you instead of straight up. That means keep this away from your house, too!

Finally, if you are on pavement, like a sidewalk or a park or something like that, make sure you don't do this anywhere near a road so the balls don't go flying onto the street. That could be really bad!

Cris Johnson

What are some others ways to experiment with this:

You can try this experiment in a lot of different ways! I was taught to do this with a basketball, but there's no reason why you can't try this with a volley ball or a bouncy ball, if that's all you have.

You just want to try it with any big ball that bounces. I don't think a bowling ball would work because they don't have any bounce to them. They just go THUD...and sometimes they go THUD on my foot. I don't go bowling very often because I'm kind of clumsy and bowling balls hurt my feet!

For the second ball, you also want a smaller ball that bounces. That's why tennis balls work so well - because they bounce!

I have also seen this done with THREE stacked balls! You would use a basketball, stack a tennis ball on top of it...and then on top of the tennis ball, stack a ping pong ball! They are super tiny and should go really high in the air!

Holding three balls like that is pretty tough for one person, so you may want to have a second person help you.

I have always seen this experiment done with two or three balls that are different sizes. I have never seen this done with two big balls. In other words, you could try stacking two basketballs on top of each other and see if the second basketball, the one on top, bounces higher than either of the two balls would bounce on their own.

I'd love to see that, so if you try that, let me know what happens!

With every way of doing this experiment, I've always said the balls, whether they are big or small, must bounce. That's why bowling balls won't work and I don't think baseballs or softballs would work very well either. They are all too hard.

And Nerf or spongey type balls would not work for this experiment either, because they don't have much bounce…in other words, they don't 'spring' back the same way that a basketball or tennis ball does when dropped.

As it turns out, that's the key that makes this entire experiment work!

The science of the experiment:

This experiment has to do with three things:

- Potential energy
- Kinetic energy
- Elastic potential energy

Let's look first at potential energy. This means that based on where an object is, it may or may not, have a lot or a little, potential energy. "Potential" basically means "it could happen."

So take that basketball and set it in the grass in your yard and look at it. If you nudge it, do you think it will roll far? If your lawn is flat, do you think your basketball could fall off a cliff and bounce or roll away? Nope.

That means when your basketball is just sitting on the grass, it doesn't have much potential energy. That means that unless we pick up the basketball and move it somewhere else, it's not going to do much. So not a lot of potential...not a lot of likelihood...for anything cool to happen.

Now pick up the basketball and hold it in your hands, arms stretched in front of you, with the ball over the grass. If your basketball drops and hits the grass, it will have more potential energy. That means more could happen with that basketball. Not much more, because the grass won't let the basketball bounce much.

Go over to the driveway. Get a piece of pizza! Hold that over the driveway! We know that the surface of the driveway is one of the things that helped that basketball bounce.

So the surface helps us determine how much potential energy an object might have.

While that pizza is held over the driveway, do you think that pizza has a lot of potential energy or just a little? Remember, the basketball bounced quite a bit more off the driveway than the grass!

Super Science

If you answered, "You're being silly! Pizza won't bounce at all!" then you're right. Because pizza doesn't bounce at all, it still doesn't have very much potential energy...but it does have SOME potential energy, because it can possibly fall down onto the ground, which would involve some energy.

So, for an object to have a lot of Potential Energy while it's held in the air over the ground, it needs to be an object that would either break apart and shatter (like an ice cube) because when something shatters, that involves a lot of energy or, like with this experiment, it needs to be something that bounces. That's why pizza wouldn't work.

We also now know that the surface the object is dropped on needs to be smooth and hard, like a driveway. That's why the ball won't bounce nearly as much on grass.

So that's a little about Potential Energy. Now let's talk about Kinetic Energy.

Kinetic Energy is the fun part: when something happens! When you let go of whatever it is that you are holding, now actual energy is being used, so the pizza will hit the grass with a 'plop,' which involves a little energy. Or when the tennis ball launches into the air off of the basketball, that involves a LOT of energy!

So we can think of Kinetic Energy as energy that is happening - the wait is over!

The third thing to talk about is the Elastic Potential Energy. In this experiment, that basically means that because a ball that bounces compresses and then releases very quickly that it has the ability or the power to use a lot of energy. That's why a bowling ball has practically zero Elastic Potential Energy: Did you ever try to squeeze a bowling ball? I have and it's no fun! It has no 'give,' so all I did was hurt my fingers.

So to add all of this together, the Potential Energy, or the ability for something cool to happen, is because the balls are held in the air over the driveway.

The Potential Energy turns into Kinetic Energy when you let the balls go. That basically just means we go from "something COULD happen" to "Now something IS happening!"

And because both the basketball and tennis ball compress, they flatten just a teeny-weeny bit. That means when the basketball unflattens, it helps push that tennis ball into the air.

Another thing - since the basketball is so much bigger than the tennis ball, it pushes the tennis ball into the air much higher than it would have gone by itself.

So, we have learned a LOT of science just by bouncing a couple of balls! And a pizza, if you were gross and really did that part.

Super Science

Don't forget to pick that pizza up off the ground.

Similar experiments:

You can see this kind of science in action in things you might already have either seen or even played: sports!

Here's just a few places you might find science like what we just learned or at least sort of like we just learned:

- In baseball, when the ball thrown by the pitcher and is hit by the batter
- In dodgeball, when the balls bounce off of players hit during the game
- In tennis when a player hits the flying tennis ball with their racket
- In billiards, when a player hits one ball and uses it to knock another ball into one of the pockets
- In ping pong (or table tennis) when a player hits the ping pong ball with their paddle

I don't know a lot about very many sports, but if you are a sports fan, there is a LOT of science in sports such as Physics, like we have talked about, as well as Mathematics (there are so many things to calculate in sports!).

So after you put this book down, look at any sports that you like and see if you can figure out where the three things we talked about (Potential Energy, Kinetic Energy, and Elastic Potential Energy) are in any sport that you like.

And if you don't care for sports, you can still look at sports that your friends or family like and point out some of this science stuff that you know about to them.

If there is something you don't like but other people around you do, finding the science in it is a really fun way to talk to people about something that they like and enjoy.

Which of the 4 sciences does this fit:

This is a Physical Science experiment because none of those balls that bounce are alive! If you dig a little deeper into the Physical Sciences you will find that this experiment is part of the Physics Sciences.

"Physics" can be defined as the study of matter and motion. It also studies things like heat, mechanics, light, electricity, sound, and more!

So Physics studies things like, how does a car start up and keep running?

Why does metal get colder or hotter quicker than wood? Physics!

How does your smart phone or iPad make sound? It's all revealed by...spaghetti! No, just kidding, it's Physics!

When you look at all the cool toys, gadgets, and gizmos that we have in life that make things easier for us or just help us to have more fun, there is probably no more useful or essential branch of science than Physics!

Super Science

Oh by the way..."essential" means we need it!

Here's something fun you can do later after you are finished reading this book...look around you and see if you can guess just how much of the things you use are here because of Physics! (Hint: It will be soooooooo much stuff!)

The website where you can see the experiments:

www.SuperheroScienceBook.com

You will need your parents to fill out the contact form to get access to the videos. Make sure you only go on the Internet with your parent's permission!

Our next experiment is going to involve another crazy optical illusion, so get ready!

Cool Science Questions:

1. What is potential energy?
2. What is kinetic energy?
3. What is elastic potential energy?

Cris Johnson

Chapter 8: Mutations

With this fun illusion, it looks like you're growing another finger! Just by putting your two fingertips together, you can create this illusion anytime, anywhere!

What is the superhero power?

If you look at this illusion like a superhero power, there have been many superheroes who have had the ability to duplicate themselves, whether it's the ability to create more versions of themselves so they have more people to fight bad guys, or simply being able to make a second or third image of themselves to fool someone, this ability has a lot of storytelling potential in comics.

Another way of looking at this power is as if it was a mutation, like maybe a scientist created a potion that he hoped would give him more power, but instead this potion caused his skin to turn purple and he grew two extra arms and two extra legs and became this crazy bad guy called Octopus Man!

Regardless of whether you look at this as an illusion or a mutation, it's still pretty cool!

Materials needed:

The beauty of this experiment is that all you need are your two hands, your two eyes and…uh…that's it.

I'm serious, that's all you need.

What are you waiting for? You can go on to the next section.

What is the experiment?

Here's what you do:

1. Pick a spot on a wall straight in front of you.
2. Make two fists…then hold out your first finger on each hand.
3. Hold your hands about three inches away from your head.
4. Turn your hands so those first fingers that you extended are facing each other.
5. Touch those extended fingertips together.

At this point, you should see what looks like a third finger between your two extended fingertips!

If you are having trouble, try moving your hands closer or further away from you until that third finger appears.

Once you have found that third finger, try pulling your hands away from each other just a little bit so that your fingertips are no longer touching, but are instead separated by half an inch or so of distance.

Now you should see that third finger floating by itself in the air, separated by a little distance from those two fingertips! This looks really wild!

Why is it a "superhero power?"

This could be considered a superhero power because people can't just decide to grow additional major body parts whenever they want. That's the mutation part.

Notice I wrote "additional MAJOR body parts." That means if you are in a horrible accident and you lose an arm, it's not going to grow back. However, we DO grow new skin, nail, and hair cells.

Your fingernails and toenails keep growing and growing. That's why you have to keep getting those nails cut! Same thing with your hair: It just keeps growing, so unless you decide to wear your hair long, you have to get it cut throughout your life.

Your skin itself is very interesting! While it varies from person to person, the top layer of skin covering your body will regenerate itself roughly every 27 days. That means your are constantly growing new skin!

Super Science

I'm just disappointed that people don't shed their skin like a snake, in one big long piece! That would be SO cool!

People also do not have the ability to cause others to see duplicates of themselves...most of the time. Let me explain!

As I mentioned earlier in this book, my favorite branches of science are the ones dealing with the mind. I help people change their lives by helping them change how they think and feel about stuff. For instance, there are people who come to me to help them stop smoking cigarettes, and they should! Cigarettes are gross!

Here is where it gets really cool...when I'm helping people change their minds about stuff, there are times where they get so relaxed, and their brains start doing really interesting things, that by just saying things to these people, I can suggest to them that they can see all kinds of weird stuff!

I do this a lot for high schools and colleges. The teachers and parents will ask me to come to their school and do a show for the kids where they think in their brains that they are on a nice warm beach, or maybe driving a cool car...and, in some cases, I'll suggest to some of these kids that they have 11 fingers, so when they count their fingers...they count to 11!

There are all kinds of things like that people like me who understand this stuff can do safely with people just for fun.

It's amazing, so the superhero ability to cause others to see extra body parts or see stuff that isn't really there is the *closest* thing in this book to actually being a real superhero power!

Where should I do this?

This experiment is so safe that you can do this practically anywhere...as long as you are not moving. If your mom or dad is driving you somewhere, or if you are on a school bus, this is not something you want to do at that time, because if the car or bus hits one good bump, you could accidentally poke yourself in the eye with one of your fingers!

Safety:

Like I mentioned in the last section, the danger with this experiment is mostly just being careful not to poke yourself in the eyes with your fingers.

Also, I don't know about you, but whenever I try to look at something where it is so close to me and in the middle of my face that I have to cross my eyes, that hurts, at least a little! This is not an experiment you want to do for long periods of time.

What are some others ways to experiment with this:

There are a couple of things you can do to change this experiment a little bit.

Super Science

First, you can try other fingers and see if it works just as well. You could also try your toes if you are 'bendy' enough. Me? I'm definitely not bendy enough. If I try to get my feet up anywhere near my face…it's going to hurt. A lot.

Second thing you can do is this: when I learned this, I was told to make sure to stare at the wall in front of you and not look directly at where your two first fingertips touch. I was told that by looking right at where those two fingertips meet would cause that third finger to "disappear."

Every book and website that I have ever read about this experiment says the same thing: that third finger will disappear if you look right at those fingers…

However, for me, if I look right at where the two index fingers touch, I can actually see that third finger better!

So the point of that story is to not just accept what the other books say without trying it for yourself.

That's one of the best things about science: if there are good, open-minded people working on something, they are going to try and get the most up-to-date information possible. A good scientist does not care about being proven right. That's so important.

A good scientist only cares about getting accurate results.

So, what that means is this: whenever possible, test something for yourself. In this case, testing whether you see that third finger is more visible when you look directly at the fingers as opposed to looking at the wall in front of you is something that is super-easy to test.

On the other hand, there are things that I'm not willing to test.

If I see one of the space shuttles being launched by NASA, and I have reason to believe they got something wrong, I'm not going to go running down to Cape Canaveral (That's in Florida where they launch stuff) and get to the building and start banging on the door, yelling, "Hey! I think you got your fuel mixture wrong! Let me in! I'm going to test it!"

But you know what? I can continue to read books, watch videos, and talk to smart people about it to see if I can make sure all the information is accurate.

Sometimes that's the best you can do.

The science of the experiment:

This experiment has been around a REALLY long time! Almost 100 years! When researching this, I found a website that mentioned a report that someone had written wayyy back in 1928! It's probably much older than that because something is very often around for a while before anyone writes about it. The paper was mentioned by a group called the American Psychological Association. It's a big organization that studies how people think.

Other times, instead of talking about a "third finger," others will talk about the Frankfurter Illusion. Do you know what a frankfurter is? Guess what!

It's another name for a sausage and is closely related to...a hot dog!

So all of that stuff earlier in the book about flying hot dogs comes back!

But it's true: in addition to being called the Third Finger Illusion this is sometimes called the Frankfurter Illusion because some people think that third finger kind of looks like a little sausage!

Me? I don't think it looks like a sausage. Mostly because when I do the illusion, I can usually still see my fingernails!

What causes you to see the third finger? Well, it is a lot like the Hole in Your Hand Illusion. It's based on the fact that since your eyes are set a few inches apart from each other, they each see a slightly different picture. When you are looking at things normally, your brain puts the images from each eye together and presents you with what you think is one whole simple image.

In this case, because your two fingers are SO close to your face, it confuses your brain, so the picture that your brain tries to put together doesn't make any sense, and you see something that is not really there: a floating extra finger!

By the way, I do understand why it is called the Third Finger Illusion: The whole experiment focuses on your first finger of each hand, and the extra finger the illusion causes.

However, since most people have ten fingers total, and this illusion makes it look like you have an eleventh finger…shouldn't this be called the Eleventh Finger Illusion?

I'm not sure who to talk to about this but I think we send emails to our congressman, senators, and mayors to ask them to talk to all of the scientists and get this changed.

What do you think: Third Finger or Eleventh Finger Illusion?

Similar experiments:

If you are reading this book straight through from beginning to end, then you know we have already done another optical illusion: It's the "Hole in Your Hand" experiment, under the "X-Ray Vision" chapter in this book.

That experiment is very similar to this one, where because our eyes see two slightly different images, our brain has to combine those images together so we "see" the world in one picture.

Later on in this book, we will be doing another optical illusion where it looks like one of our hands is passing through the other!

Which of the 4 sciences does this fit:

Just like with the Hole in Your Hand experiment, this could be considered a Life Sciences experiment because it's involving you and me, living people! Or it could also fall into the Social Sciences branch because we are dealing with psychology and sciences like that.

What I find really interesting is that, in many cases, scientists themselves don't always agree on how a science experiment should be classified! In the case of many optical illusions, there is debate among scientists over how to classify them!

YOU can do even more reading and make a decision for yourself where experiments like this should be grouped.

The website where you can see the experiments:

www.SuperheroScienceBook.com

You will need your parents to fill out the contact form to get access to the videos. Make sure you only go on the Internet with your parent's permission!

Our next experiment will involve something that is very EASY to do and is fun to have contests with your friends! You're going to love this!

Cris Johnson

Cool Science Questions:

1. What is another name for the Third Finger Illusion?
2. How long has this experiment been around?

Super Science

Chapter 9: Creating Evil Science Serpents!

This experiment is a little different because it doesn't look much like a superhero power to me but more like something a mad scientist would do!

That reminds me of something: As I go through all of these experiments in this book, I mention what 'superhero power' they remind me of, but that is just my imagination. You might look at some of these experiments and think they remind you of something totally different and that's great!

What is the superhero power?

So, this experiment is all about creating evil serpents. Maybe there is a crazy scientist who wants to take over the world and he (or she...girls can be evil scientists too) created these weird serpent things to help them as their loyal but evil minions.

Here's the thing...in this experiment, they are not really evil and they are not really serpents. They are bubble snakes!

Materials needed:

For this experiment, you will need just a few things:

- A 16oz empty water or soda bottle
- A bowl or large plastic container
- Dish soap: Dawn dish detergent seems to work the best
- A long sock
- A cup of water

What is the experiment?

To prepare, you'll need to have your mom or dad help you by cutting off the bottom of the bottle with a pair of scissors or some other cutting utensil.

Pull the sock over the bottle and keep pulling and working the bottle into the sock until the toe of the sock is tight against the opening that was cut in the bottom of the bottle.

If you want, you can also just use a washcloth or some other smaller piece of fabric that feels the same as the sock. You can hold that smaller piece of cloth to the bottom of the bottle by using a rubber band.

When I did this experiment, I just used the sock.

Super Science

Whether you use the cloth or the sock, when you finish, make sure the small opening for the bottle, where the bottle cap would go, is open. That's where you are going to blow.

Next, pour the cup of water into the bowl or plastic container. Squirt 2-3 tablespoons of the dish soap into the cup. Mix it very well.

Take the bottle and dip the end where the bottom was cut off into the soapy water. Let some of the liquid soak into the cloth. Take it out of the soapy mixture and shake off the excess.

Now it's time to make bubble snakes!

Gently blow into the bottle through the open top of the bottle, where you would normally drink from. In just a few moments, you'll start seeing bubbles coming out of the bottom! You might get a big ball of bubbles or you might get a snake - it seems kind of random.

Keep blowing and see how long your snake gets before breaking!

Why is it a "superhero power?"

This could be considered a superhero power (or super villain power) if the snakes were living because people do not have the ability to just create living things out of stuff from around the house.

However, scientists are working on all kinds of incredible-sounding experiments, like cloning!

You might be wondering, 'what is cloning?'

Cloning is the idea that scientists can take a few cells from one animal and create a brand new animal that is exactly the same as the first animal in every way!

When I was a kid growing up, this sounded like something totally made up, from science fiction books, comic books, or movies. The fact that scientists have actually cloned animals a bunch of times just blows my mind!

The first successfully cloned animal was a sheep named Dolly, many years ago. Since Dolly, other animals have been cloned, including baby monkey clones!

Many scientists now believe that cloning humans is very possible. Many people support the idea, but many others do not.

Cloning humans might be an interesting thing for you to research after you are finished with this book!

Where should I do this?

This experiment is safe enough to do almost anywhere, indoors or out, but since a lot of the bubbles could break off of your snake and fall to the floor and make a sticky mess, this is one experiment you should do outside!

Super Science

When you go outside, it is best to find a place that is not very windy. The more wind you have, the more likely your bubble snakes are going to break before they get very long.

Safety:

As I mentioned earlier, this is actually a really safe experiment, but there are just a few things to remember to stay safe:

1. Do the experiment in a place where if the soapy bubbles fall to the ground, it is not going to make the surface slippery to walk on.
2. Always remember to blow OUT with your breath when making bubble snakes. Don't suck air in because you could wind up getting dish soap in your mouth and that would make you sick in your stomach...plus it would just be gross!
3. Make sure you have your mom or dad help you cut the bottom off that bottle. It's real easy to get hurt using sharp objects.

I'm hoping kids of all ages read this book, so I'm being really careful to talk about safety. If you are an older kid thinking "Geez, there's no way I'd be silly enough to cut myself with scissors," I have a story for you!

I have mentioned a few times in this book that when I visit a school to do an assembly program, I often do a lot of magic in my shows. One of my tricks is taking a pair of scissors, cutting a long piece of rope in half, and magically causing it to become one rope again.

It's a trick I've done about a million times (Okay, not really a million times...my mom has told me a billion times to stop exaggerating), and because I've done it so many times, I don't really 'think about it' when I'm doing it.

So one day I was doing a magic show outside and I was doing the rope trick...and I cut one of my fingers REALLY bad with the scissors. It took a really long time to heal. Years later I still have a scar.

Even grownups can hurt themselves doing super-simple things, so be careful.

The last thing to be careful: Use a CLEAN sock! You don't want to have your nose near a stinky sock while you are blowing snake bubbles!

What are some others ways to experiment with this:

There are all kinds of things you can do to change this experiment. Here's just a few:

- One thing you can do to make your bubble snakes prettier is with food coloring! After you dip the sock end of the bottle into the soapy water, squirt a few drops of one or two different colors onto the sock. As you blow the bubble snake out, it will come out with streaks of those colors!

- You can try letting the soapy water sit for several hours before you start blowing bubble snakes. Some scientists say that letting the soapy water sit for a while will help the bubbles become stronger so they don't pop as fast.
- Try adding some glycerin to the soapy mixture. Scientists say that will also make the bubbles stronger. Glycerin is a simple liquid that can be found at pharmacies. Some scientists say you can also use something called Karo syrup instead of the glycerin to strengthen the bubbles. Karo syrup can be found at grocery stores.
- This is the most fun: have contests with your friends! See who can make the longest bubble snakes! When you watch the video of me doing it, I made a bubble snake that was about 5 feet long!

The science of the experiment:

The thing that makes this experiment so much fun is the bubbles! Bubbles are awesome.

A bubble looks like it's one very thin delicate layer in a round shape, but it is actually made up of three layers: there is a layer of water that is contained between two layers of soapy molecules.

Most people think of bubbles as having air in the center, but bubbles can be created using different gases, such as carbon dioxide.

Cris Johnson

If you don't know what carbon dioxide is, try this:

- Take a deep breath in. You're taking in a gas. That's the oxygen that keeps you alive.
- Now breath out. That's the carbon dioxide. That's a different gas, one that is bad for humans, which is why we breath it out.
- With plants, it's the opposite: when plants 'breathe' in, they are taking in carbon dioxide. When they 'breathe' out, they are giving off oxygen. So the more plants we have, the more air we have to breathe. Plants are awesome!

Did you know why bubbles are round? It's because it is the easiest shape for bubbles to be without popping. The round shape reduces how much surface tension that there is. The greater the tension, the more likely a bubble is going to pop.

When a bubble is round, it means that the pressure on the wall of the bubble is equal all the way around, so there are no weak points, so the bubble has more of a chance to last longer.

Similar experiments:

There are all kinds of things you can do with bubbles! One of the easiest things you can do is get some bubble wands and see how long you can make a single bubble!

Super Science

With many bubble wands, they are simply pieces of plastic with a round shape and a handle. You dip the bubble wand into the soapy mixture in a bottle, take the wand back out, and blow bubbles.

Some bubble wands are pieces of string tied into a circle with a couple of plastic handles on them. These devices allow you to make bubbles with a wider circumference, so you have a bubble that is a lot wider!

Circumference is just the imaginary line that goes around a bubble. Imagine if you could take a tape measure and wrap it around a basketball to find out how big it is.

So you can take these special string bubble wands and make really giant bubbles! This is definitely something you want to do outside.

When I do a live version of Superhero Science in a school, I use a great big bubble wand that looks sort of like a hula hoop and I make a giant bubble right around a kid, so we create a Kid In A Bubble!

Trust me, that is really awesome!

Which of the 4 sciences does this fit:

Bubbles are definitely in the Physical Sciences. We might think that bubbles just look pretty, but scientists are working with bubbles to learn more useful things for Engineering.

Engineering is the study of building things, all kinds of things, using science. If some engineers have to build a building, there's a bunch of stuff they have to think about, like making sure strong winds and other weather won't knock the building down. They have to make sure the building is safe for people to be in. All kinds of things like that.

One of the things scientists are doing with bubbles is creating them in a weightless environment to see how they move, explode, and things like that.

A 'weightless environment' means there is no gravity so anything in it would float! (If you remember when I explained gravity earlier, it's what holds you to the surface of the Earth!)

Scientists are making bubbles in a weightless environment so that the bubbles' shape is not being changed by the gravity, so they are the most perfectly round bubbles possible.

So these people are using science to study two of my favorite things: bubbles and floating in the air!

I don't know about you but to me that sounds like fun!

The website where you can see the experiments:

www.SuperheroScienceBook.com

Super Science

You will need your parents to fill out the contact form to get access to the videos. Make sure you only go on the Internet with your parent's permission!

Our next experiment is not messy at all and you can either do it alone by yourself or with some friends watching. You can even do this for friends over the Internet!

Cool Science Questions:

1. What are the three layers of a bubble?
2. What is cloning?
3. What does circumference mean?

Cris Johnson

Chapter 10: Passing Through Solid Objects!

This is an experiment you can do pretty much anytime, anywhere. The first time I saw this, I thought I was watching a special effect from a movie!

What is the superhero power?

This superhero power is the ability to physically go through solid objects!

Many superheroes have had this ability to pass right through walls in order to get the bad guys or escape from being trapped.

Depending on the superhero, the person might turn to smoke or gas and pass through the wall that way. Other characters in movies and TV shows become a sort of liquid and squeeze through small spaces, like between the bars of a jail.

Super Science

Still other heroes 'vibrate' so quickly that they are able to pass their molecules through solid objects by vibrating between the molecules of a solid object.

However they do it, it's a pretty awesome power!

Materials needed:

All you really need for this are your own two hands! If you are doing this by yourself, you also need a mirror or some other way for you to see what you are doing.

If you are doing this for your friends (either live with them right in front of you or over the internet), then you don't need the mirror.

What is the experiment?

This is a fun optical illusion! Here's what you do:

1. Hold up your left hand with your palm facing away from you.
2. Spread all five fingers far apart from each other so none of them are anywhere near touching.
3. Take your right hand and spread those fingers as far apart from each other as possible.
4. Put your right hand behind your left hand. Your right hand's palm should be pressed against the back of your left hand.
5. Close your right fingers over the spaces between your left fingers, so your right fingers are holding

onto your left hand. It should look like your right hand is hanging onto your left hand from behind.

6. Open your right fingers and pull your right hand away from your left hand just an inch or so, leaving your right fingers wide open.
7. Press your right palm against the back of your left hand again, while at the same time closing your right fingers around your left hand again.
8. Repeat this action 3 or 4 times.
9. Then quickly close the fingers of your left hand into a fist and open your right hand's fingers again, but this time, keep your right hand itself still, so the whole hand does not move, just the right fingers open.
10. As soon as your left hand is in a fist and your right fingers are open, move your closed left hand away from the right hand very quickly.

Reading all of that might seem like it is very hard, but a little practice is all you need. Also, watching the video of me doing it will help too.

When you do it right, it looks like one hand is passing through the other. It's SO cool! You can fool yourself doing it!

When you do this, you do want to be pretty close to the mirror or your friend's face or your computer. You don't want to be so close you could touch someone as that's too close.

Super Science

You also don't want to be so far away from someone that they can't see what is supposed to be happening.

For instance, you don't want to do this on one end of a football field while your friend is at the other end. If you try it that way, your friend will probably shout, "Hey, I can't tell what you are doing. Can you throw me the football?"

Watching my video will help give you an idea of how far away you should do this.

Why is it a "superhero power?"

Solid objects passing through solid objects is impossible. One of the basic laws of physics states that two objects cannot occupy the same space at the same time. And if one thing is passing through another, then at least for a split second, they would be in the same space.

However, in life we often see things where it sort of looks like things are passing through each other.

Think of when it's really cold outside in the winter (if you live in an area where there is snow). People who are cold will often wrap a scarf around their face, covering their nose and mouth. Or they will wear a ski mask that leaves their eyes and mouth uncovered, but keeps their nose from being exposed to the cold air.

These people (and maybe you) are still able to breathe, because while the air (which is a gas) does not actually pass through the cloth (which is a solid), the air molecules do pass between and around the cloth molecules.

A cloth is not airtight, but something like a plastic bag is which is why parents are always being warned not to leave empty plastic bags around the house where very little kids or babies could pull them over their faces. They just don't know any better because they haven't gone to school and learned stuff like you have.

It's the same thing with water and other liquids. Because they flow, they can appear to pass through solid objects if the solid objects are the kind where they are not watertight or airtight, meaning there is space between molecules to allow the liquid through.

What all of this means is that while passing solid objects through solid objects is impossible, liquids and gases can appear to go through solids even though they are really going around or between the solids.

So, don't try to use your dirty sock to carry around water with you to drink. It won't be in the sock for very long!

Where should I do this?

This is really great because you can do this anywhere, indoors or out! The only thing is, you either need some people to watch you doing this so they can experience it, or you need a mirror so you can see it yourself.

Doing this by yourself without a mirror means you are just playing with your hands, and while there is nothing wrong with that, it would be kind of boring!

Safety:

This is pretty much safe to do anywhere.

I say "pretty much" because you wouldn't want to do this while riding your bike because you might lose your balance and fall down. You also don't want to do this to your mom or dad while they are driving.

Flinging your hands in their face might surprise them and they might scream "Alien! You must be a space alien if you are able to pass one hand through the other!" and they might crash the car because they are so scared.

What are some others ways to experiment with this:

You might try this at different speeds to see which way is more magical looking. On the video of me performing this, you'll see me doing this a couple of different ways.

You can also try folding your right fingers over your left fingers more times to see if doing it more times makes the illusion of your hand passing through the other look even more amazing!

The science of the experiment:

The whole thing that makes this experiment work is an idea called "Conditioning." Conditioning is the idea that if we do, hear, smell, taste, or see something enough times then we will react in a certain way.

In this case, if you are doing this into a mirror, you watched your left hand stay open and still while your right hand was opening and closing over and over again. Your brain got used to seeing that same motion, so when everything suddenly changed, your brain was used to the old motions, so it was tricked for a moment into thinking it saw one of your hands pass right through the other!

We see "conditioning" everywhere. Let's pretend you visit your grandma every weekend. Every time you visit your grandma, she plays a certain song for you at her house. Now let's pretend you go on a trip really far away and you are in a store or restaurant and you hear that same song. As soon as you hear it, you are probably going to think about your grandma.

Conditioning is also things we do automatically, kind of without thinking about it or really trying too hard. When you learned how to tie your shoelaces, it might have seemed hard to do at first. Then you practiced and did it a bunch of times until it became so easy that you just kind of tie your shoes without thinking about all the ways your fingers and hands have to move. You just decide to tie your shoes and your hands and fingers know what to do.

Super Science

Think about your life and all the things you do and see if you can name at least 10 things you do without thinking about them.

Conditioning has been around a really long time and if you think about it, it saves us a lot of time. Because we do so much stuff without really thinking about every little step, we can do more stuff faster every day.

Here's another way to understand conditioning: Say the alphabet out loud. Go ahead. Right now. If anyone asks you why you are saying the alphabet out loud, just tell them this weird guy who wrote a book told you to do it.

So if you actually said the alphabet out loud, what happened? You said each letter pretty fast. That is because when you were really young, your teachers had you say the alphabet out loud over and over again until you could say it without really thinking about it. Your brain just kinda let you do it.

Now try this part: Say the alphabet again...but this time, skip every other letter! So you would say, "A...C...E..."

Try it, at least for a few more letters!

Okay, you don't really have to do the whole alphabet that way. But if you really tried to do it and just didn't read my words, I'll bet it was harder for you and you had to stop and think about each letter.

Cris Johnson

It really should be easier, right? I mean, you only had to say every other letter, so you only had to say half of the alphabet, so it was only half the work, right?

Nope. For most people, it's wayyy harder saying the alphabet every other letter. Why? They were never asked by teachers when they were really young to say the alphabet like that. So their brain never got used to doing that.

That's the difference between doing something without conditioning and doing something you have been conditioned to do. It's really neat!

You can see this in animals, too. My dogs love their food. They look forward to it every day. So do my cats. But this is interesting: Every day I take a little can of cat food out for my cats and I take out a can for my dogs. When I open the cat food can, my dogs don't do anything. But when I open the dog food can, they get really excited and start barking because they know their dinner is coming soon.

I don't know if the cans sound different when they are opened, because to me it sounds exactly the same if I open the cat food or the dog food can. Or maybe the dogs can smell their dinner as soon as the dog food can is opened. Dogs have really good noses. All I know is when that can is opened, my dogs go crazy!

This reminds me of a really important scientist named Ivan Pavlov. In the 1890's he started doing experiments with dogs. He wanted to see how much dogs would salivate, or drool, when a meat-flavored powder was placed in the mouths of dogs. His assistant would bring the powder to the dogs every day.

What Pavlov discovered is that pretty soon the dogs would salivate when they heard the footsteps of the assistant getting close to them. They had been conditioned to learn that when the assistant came, they would get their tasty treat, so they would start salivating even before they actually got the treat.

So not only do humans get conditioned, so do other animals.

Dogs are so awesome at conditioning (and dogs are just awesome anyway) so when you visit the website with this book and you watch the video for this experiment, you'll see me working with my dog Stanlee to show you conditioning.

Similar experiments:

Although their optical illusions happen because of your eyes being separated from each other by a small distance (which is different from why this experiment works), you can still do other optical illusions right in this book!

Those chapters are called "X-Ray Vision" and "Mutations."

If you have been reading this book in order, right through, you might be thinking, "Why is he telling me about experiments I've already read about? Is he getting old and forgetting stuff? Will he remember to even finish this book?"

Yes, if you have been reading this book right straight through, then you have already read those other chapters. But since this is a science book and not a story, it means some people reading this might skip around and read their favorite parts first.

So yes, I absolutely will finish this book...although that would be kind of a surprise to read a book, and get to the end, but the book does not have an end.

I would never end a book without finishing it. However, it might be kind of funny to end this section of this chapter right in the middle of

Which of the 4 sciences does this fit:

This experiment is in the Social Sciences field and it's in the Psychology field. Psychology, once again, is studying how people think and behave.

This optical illusion trick is different from the other two optical illusions in this book because the other two were caused by your two eyes seeing two slightly different things and your brain squishing those two pictures together so you saw something that was not real.

For this experiment, since the hands are seen farther away, both eyes see the same thing. The illusion comes from your brain getting used to seeing the same movement over and over. Your brain gets used to this movement, so when the movement suddenly changes, your brain is surprised by this and thinks it's seeing something it is not.

Working with the brain is so cool! These are the kinds of things I do all the time as a hypnotist. When I do fun shows at high schools and colleges, I help people use their brains to see and experience all kinds of things that are not really happening.

The website where you can see the experiments:

www.SuperheroScienceBook.com

You will need your parents to fill out the contact form to get access to the videos. Make sure you only go on the Internet with your parent's permission!

In the next chapter, we are going to do an experiment that looks like one of the most popular superhero powers!

Cool Science Questions:

1. What is conditioning?
2. Who was Ivan Pavlov?
3. What animal did Pavlov work with?

Cris Johnson

Chapter 11: Invulnerability

When it comes to fighting bad guys, this superhero power is one of the most useful!

What is the superhero power?

Here we are going to talk about invulnerability, which is the superhero ability to not be physically injured when struck by a bad guy's punch, gun, foot, bomb, club, spear, bullet, karate chop, or smelly socks.

In short, this means some superheroes can take a LOT of punishment to their bodies but not get hurt.

With some superheroes, this is a true superhero power, where their bodies are strong enough to withstand attacks. With other superheroes, they protect themselves with armor or some other physical protection that keeps their bodies safe.

Still other superheroes can use magic or spells to protect themselves, so 'invulnerability' can take a lot of different forms in comic books and superhero TV shows and movies.

Materials needed:

This experiment only needs a few things:

Super Science

- A Ziplock (or other brand) one quart freezer bag
- Sharpened pencils
- Water to fill the bag

That's all you need! Be sure that your pencils are REALLY sharp. That way, this experiment will work better.

What is the experiment?

Okay, here's what you do: First, go some place where if water spills, it won't make a mess. You can do this experiment over a sink or outside, for example.

Then, you are going to fill up the Ziplock bag with water and then seal it. Here's the deal: you don't need to fill up the bag completely, as that might make the bag hard to hold onto for the experiment.

When I do this experiment, I usually fill up the bag a little more than half way.

Now comes the FUN part! You are going to take one of those sharpened pencils and just push it into the side of the bag: Once the point of the pencil has gone into the bag, keep pushing until the point end of the pencil comes out the other side and then STOP!

You don't want to keep pushing the pencil all the way through the other side of the bag and out as that would expose the holes you created and then water would come rushing out!

Here's what you do next: push another pencil into the bag! Then another! I usually keep going until I have 6 or 7 pencils stuck in the bag. Most of the time, absolutely ZERO water leaks out so this experiment looks completely impossible!

This is important: to end the experiment, you have to pull the pencils out of the bag, which will cause the water to spill out of the holes once the pencils are not there. This means you have to do this somewhere safe!

Why is it a "superhero power?"

Invulnerability is considered a superhero power because in real life, a human's skin is actually pretty easy to damage. I know this because even when I exercise and get good and strong, if one of my cats decides to scratch me when I take them to the veterinarian, then I still shout "Ow, that hurt! You evil cat!" as soon as I feel those claws.

In nature, you see certain animals have evolved natural forms of protection. For instance, the skin of a rhino or an elephant is much tougher than a human's skin. This makes sense because rhinos and elephants can't wear sweaters or jackets to keep themselves warm when it gets cold out. (I tried putting my jacket on a rhino. It just didn't fit his style.)

Other animals, like turtles and tortoises, grow their own armor in the form of their shells which help to keep them pretty safe.

And then there are hermit crabs: These little guys go find sea shells on beaches and then carry them around wherever they go. If they grow and get too big for one shell, then they look for a new bigger one. It's sort like you - most kids grow quite a bit so your mom or dad is constantly getting you bigger clothes.

Where should I do this?

Like I wrote earlier, this is not a dangerous experiment and if you leave the pencils in the bag, you usually won't get any water leaking out. (Once when I did it, I had a little water leaking out, but I think one of my pencils may not have been sharp enough.)

But since we are working with water that could spill, and because a good scientist is always careful, this is another experiment to do outside. However, if it is too cold to do experiments outside where you live or if you don't have a yard where you can do this outside, you can easily do this over a sink or even in the bathtub! A quart of water isn't very much but it's enough for your parents to yell, "What's going on here? That science book you are reading is going to destroy our house!" so be careful.

Safety:

Other than possibly spilling some water, there is only one other thing to be careful doing with this experiment but it is important: be careful with those super-sharp pencils.

Cris Johnson

When I was in school, I knew of a lot of kids who accidentally poked themselves with pencils that were really sharp...Okay, it was me who did it when I was in school. I'm sorry I did it but I really did not think about how sharp those pencils were!

So be careful with sharp pencils.

Also, to reduce the possibility of letting any water leak out, use completely round pencils instead of pencils that have flat edges on them. On the video of me doing this experiment, a little water leaked out for me and I think I might have been using pencils with flat edges.

What are some others ways to experiment with this:

If you want the experiment to also sort of look like an art project, you can try adding food coloring to the bag to make the water look neat. And you can use different colored pencils because having all different colored pencils sticking out of the bag would look really cool.

You can also experiment by filling the bag up completely with water. Does that make the pencils go in any easier? Does water leak out easier because more water in the bag means more pressure? It would be fun to find out!

You can also have contests with your friends! Each of you can take turns pushing pencils into the bag. Whoever pushes the last pencil into the bag without it leaking would win!

Super Science

In fact, it is sort of like a super-weird game of Jenga, only with the potential for a soggier ending to the game!

Another contest idea: each of you has a bag full of water and a bunch of pencils. So who can get their pencils into their bags the fastest without any leaking!

The science of the experiment:

This experiment works because the plastic bag is made up of polymers. Polymers are a kind of substance where all of the molecules are mostly or entirely (depending on the substance) made of the exact same kinds of molecules.

Because those molecules are all the same, it's easy to mess with them. And the polymers in the plastic bag are very flexible, so the pencils push around or through the molecules. The flexible molecules then squish back around the pencil, sealing the bag and keeping the water from leaking out.

Polymers are awesome, and you can find polymers in all kinds of things around your house.

For instance, if you have a baby brother or sister, grab one of their diapers! You can use the diaper to change the water-filled bag experiment so you can pull the pencils out WITHOUT water leaking everywhere!

You might be asking, "How? That sounds like magic!"

Cris Johnson

Inside each diaper is some powder that looks sort of like flour or sugar but it's actually a polymer powder! When the diapers get wetness in them (uh, yeah, that means when babies do their business), the powder inside the diaper absorbs all the liquid. Then the polymer and liquid mixture turns into a sort of jelly-like substance!

What's happening is the polymer powder absorbs the liquid and not only does it hold onto the liquid, but each 'grain' of the powder squishes up to the others and they all stick together so the liquid doesn't go anywhere.

Yup, science helps your baby brother or sister stay dry!

So, if you carefully open the diaper (it might be best and safest to have your mom or dad help) and shake the powder inside of the diaper into your water-filled Ziplock bag full of water and pencils, the water will turn into that jelly-like substance and you can pull the pencils out without spilling any water!

Be careful: depending on how much water you have in the Ziplock bag, you might need to use 2 or 3 diapers.

Also, if an animal or your baby brother or sister were to eat any of that powder, it might be REALLY bad, so be super-careful to make sure all of the powder is thrown safely into a garbage can when you are done.

Super-important last-minute science tip: Don't put any of the powder in a kitchen or bathroom sink, and don't flush it down the toilet! Trust me, because the polymer powder absorbs liquid, a whole lotta bad stuff could happen!

Similar experiments:

I started looking around at other science experiments online and found one where you used gelatin, a natural polymer, and a few other ingredients to make wiggly worms or bugs! You can ask your teacher or parents to look that one up for you online.

There's another experiment involving making what looks like snow! It uses a special polymer product called Insta-Snow. When you add water to it, it puffs up and looks just like fluffy snow, even though it's not cold and it won't melt! I have used it and it is super-neat! You can ask your parents to get Insta-Snow online from Amazon. Some magic shops have it too, as many magicians use it in their Christmas shows!

Which of the 4 sciences does this fit:

Polymer experiments fall into the Physical Sciences category, and specifically, polymers fall into the field of Chemistry because scientists have used chemistry techniques to convert polymers into different forms to help people in their lives.

We use SO MANY polymer products in our daily lives that it's incredible! Polymers are in garbage bags, water bottles, tires and more! They are everywhere, so they are a good thing to learn about.

There was a guy named Hermann Staudinger who was a chemistry scientist and did a lot of experiments that helped us learn how to use both synthetic (that means scientists created it) and natural (uh...that means nature created it) polymers. He did so much important work in science that in 1953 he was awarded the Nobel Prize in Chemistry. In case you did not know, winning the Nobel Prize is a really big deal and it means you did something really important and helpful to people everywhere!

Getting back to the different kinds of polymers: The stuff we use a lot in products (like garbage bags, diapers, water bottles...you know, the stuff I just mentioned a few paragraphs ago) are examples of synthetic polymers. That means we have to make them in factories.

You're never going to see a farmer watering his or her land as they grow the plastic for water bottles!

However, natural polymers are things we do see in nature - stuff like rubber and wood. I knew about wood, but I was amazed to find out that rubber was a natural polymer, meaning you can find it in some form in nature!

Here's the crazy thing: Natural rubber is obtained from a rubber tree (makes sense) but the part that blew my mind was that people get natural rubber from these trees in the form of a white liquid!

So if you have a tire or something else made of natural rubber. It started off as something that flowed like milk, water, or orange juice. That's amazing! In fact, a guy by the name of Charles Goodyear did a lot of work with rubber and improving its strength and durability.

So if you see that your parents' car has Goodyear tires, you can tell them that Charles Goodyear developed ways to make car tires stronger!

Always remember this: science is all around us, everywhere!

The website where you can see the experiments:

www.SuperheroScienceBook.com

You will need your parents to fill out the contact form to get access to the videos. Make sure you only go on the Internet with your parent's permission!

In our next experiment, you are going to learn how to do something that will amaze all your friends and family, and it's one experiment where you can combine your creativity and artistic ability with science!

Cool Science Questions:

1. What are polymers?
2. What are some of the uses of polymers?
3. Who was Hermann Staudinger?

Cris Johnson

Chapter 12: Shrinking

This is one of the experiments I was most excited about for this book! It's something that is very easy to do and can be made to look exciting, funny, silly, and impossible!

What is the superhero power?

This superhero power is the ability to shrink! Some superheroes need to shrink in order to get into very tiny spaces to fix machines, hide from bad guys, or get out of being trapped.

It's also one of the most impossible superhero powers ever!

Materials needed:

For this superhero power, you only need a couple of things:

- At least 2 people, or 3, depending on the picture you create

- A camera or cell phone to take a picture

That's it! That's all you need! Of course, if you get creative and think up some very clever pictures, you may need other props.

If you watch the video with this experiment on the website, you'll see that I included a great deal of pictures that had props, weird locations, and very funny staging to make it look as though people were really tiny, or regular sized people were drinking from giant cans of Pepsi, or were stepping on cars, all kinds of things!

What is the experiment?

This experiment involves posing people to look like one person is really tiny while another is either regular size or giant sized!

The way this works is simple: One person stands closer to the camera while a second person stands farther away from the camera.

I've seen a lot of pictures of this and taken a lot of pictures like this, so to get you started I will explain how I made it look like I was super little and my mother-in-law was holding me in her hand.

It's started by getting a couple of people to help you. You need two people for this. You are the person who is going to hold the camera or phone. The other two people are actually going to be in the picture.

Ask the person who is going to be the one who looks really teeny to walk away from you and the third person. Have them walk so far that when you look at them, they look small enough to put in your pocket. Have that person just stand there and wait.

Ask the second person to stand profile to the camera. Have this person turn to face the person who is far away, so if they were closer together, they could look at each other. Standing profile means you turn to your side so someone looking at you can only see your right or left side, depending which way you are facing.

Ask the person who is closest to you to hold up their hand, with their palm toward the sky. Now you look through the camera.

You are going to need to ask this person to move and turn, and adjust their hand up or down until it looks like their hand is holding the person really far away. Then you take your picture.

It's super-simple, but the key to making the picture look good is having the third person (the one holding the phone or camera) look at both people and make sure it looks like they are really interacting and touching each other.

Super Science

With one of the pictures that I did, it looks like I am super-little and my mother-in-law is holding me in her hand. The second picture we did looks like she is dangling me from one of my arms.

This looks really cool, but my wife, who was holding the camera, had to "line up the shot," which means that she was the one who made sure that when it was supposed to look like my mother-in-law was holding me by the hand, that both of our hands were in the right position because we were both standing about 30 feet away from each other!

It's the same thing with the other picture: my wife had to make sure the camera was in the right place so it would look like my mother-in-law was actually holding me in her hand, even though I was once again standing 30 feet away from her.

Here's a tip: if you do this picture, ask the person closest to you to move around and adjust their arm and stuff. Trying to get the person who is really far away to adjust their position is going to be tougher because they are far away. You'd have to yell at them so they hear you.

But yelling at someone is hard on your vocal cords. And the person you are yelling at might get mad because you keep yelling at them. So have the person who is really far away just stay in one spot and wait. You can talk to the person who is closer to you a lot easier because they are so much closer.

Cris Johnson

Why is it a "superhero power?"

The ability to shrink from a regular sized human down to a few inches high is completely impossible. People do not have the ability to do that.

However, people do in fact shrink throughout their lives. Some studies report that as males get older, they might lose an inch or more of height between the ages of 30 and 70.

Women might lose up to two inches of height during those same years.

So what's going on?

As people get older, their bones start to get more and more brittle. It's like an engine in a car wearing out: Things just break down.

On top of that, gravity (remember the earlier stuff in this book about gravity?) is constantly pulling on us, so that helps to wear down our bones too.

When people are in space, they don't have the gravity pulling on their bones and bodies like on Earth. Unfortunately, because people are weightless in space, their muscles don't get a lot of work.

That might sound kind of cool, but if your muscles don't get work, they start to break down and get weaker and weaker, so to be healthy and strong, we need gravity to keep us working.

When our muscles don't get a lot of work and start to get weaker, this is called "muscle atrophy," and it's so harmful to us which is why we have gym and physical education classes in school.

It's also where the saying "Use it or lose it" comes from!

Where should I do this?

You will probably need to do this outside, at least for the camera shots I was doing because you need the two people in the picture to be so far apart from each other. I guess you could do it if you were in a building that is big enough, like maybe a mall, but the people in charge at the mall really do not want people taking pictures at their mall.

Safety:

This is a completely safe experiment! In fact, I couldn't think of any way a person could get hurt doing this unless you asked one of the people in your picture to bend their body in a really uncomfortable position or if while you were taking your picture, one of the people stood in a road or something like that.

So, stay away from roads where cars and trucks go while you are taking your pictures. Of course, you are probably reading this book and shouting at me, "Dude! You shouldn't hang around on a road even if you are not taking a picture!"

Yes, you are totally right!

What are some others ways to experiment with this:

This is one of those experiments where even if you are doing the same basic thing, which is making people or objects look like they are bigger or smaller than they are in real life, there are so many different pictures that you could take, you could come up with a hundred different ideas every day and in a year and you would still not run out of things you could do for really fun pictures!

Just look on the internet with your parents to get some ideas. And the video on my site with this experiment has some really awesome ideas too!

The science of the experiment:

These 'shrinking people' pictures are possible because of something called 'forced perspective.' As I wrote about before, that just means that objects and people look smaller when they are farther away or look bigger when they are closer. So that means that by playing around objects and positions, we can make cool pictures like we have talked about before.

Super Science

Similar experiments:

Guess what? This is one more optical illusion! This one doesn't mess with your eyes like some of our other experiments. Some of the other experiments in this book messed with your eyes because your two eyes were apart from each other by a few inches so each eye saw something slightly different, and your brain has to try and smush those two images together into something you can see and understand.

This experiment is an optical illusion but it works differently because the objects (in this case, people) are far enough away that your eyes see the same thing.

If you are just skipping around in this book and reading your favorite parts, you'll see other optical illusions in this book. They are experiments based on the following superhero powers:

- X-Ray Vision
- Mutations
- Solid Objects Passing Through Each Other

Which of the 4 sciences does this fit:

Like the other optical illusions in this book, this falls under Social Sciences. It also falls under Psychology (my favorite branch of science) because it's all about how people see stuff or what they think they are seeing (or hearing, smelling, tasting, touching...), because we are talking about an experiment that deals with our "perceptions."

Perceptions is a special part of Psychology because it means we see, hear, or feel stuff around us that might not actually be happening.

You might be thinking, "What do you mean we might be seeing or hearing stuff that might not really be happening? Are our eyeballs lying to us?"

Not exactly.

Sometimes our brains might get it wrong.

Here's something you might have done: have you ever looked up at the sky, watched white fluffy clouds, and seen certain shapes or even animals in the clouds?

That is because part of your brain is always on the lookout for familiar stuff, things it has seen or felt or heard before. So, when your brain sees, feels, hears (tastes and smells, too) something brand new, it tries to compare it to something it encountered before.

In the example of those clouds, if you ever saw Frosty the Snowman in a cloud, it was never really Frosty, but it sort of looked like him.

Listen, I KNOW it was not the real Frosty in the sky with the clouds. I KNOW it because the real Frosty isn't that big. If Frosty really was big enough to be seen in the sky miles

Super Science

and miles away, do you have ANY idea how freaky big that Frosty would be if you got up close to him? He'd be huge! And scary.

The reason I bring all of this up is because sometimes what we think we are seeing is not really what is going on in the real world.

So this experiment takes advantage of the fact that your brain is constantly trying to figure stuff out and sometimes it gets it wrong. And for this experiment, that means we can have a lot of fun!

The website where you can see the experiments:

www.SuperheroScienceBook.com

You will need your parents to fill out the contact form to get access to the videos. Make sure you only go on the Internet with your parent's permission!

For our next experiment, we are going to look into the future! Or at least make it seem that way.

Cool Science Questions:

1. What is forced perspective?
2. What happens when your muscles atrophy?
3. What are your perceptions?

Cris Johnson

Chapter 13: Predicting the Future

What is the superhero power?

With this superhero power, we are talking about predicting the future! That means that by using some kind of special ability, a hero would know what was going to happen the next day, next month, next year, or even longer, depending on the story being told.

Some superheroes have this ability just as part of their abilities and they just 'know' what is going to happen in the future, similar to how some superheroes are really strong or have the ability to fly.

Other superheroes have the ability but it's kind of confusing. Maybe they have a dream about the future and during this dream they see:

- a guy wearing red shoes running
- a checkers board
- a little kid eating tacos

They then have to figure out what those things together mean for the future. In other words, the hero has to guess what is going to happen based on those images.

By the way, with the things I mentioned, do you know what is going to happen based on those three things I mentioned?

Think about it for a second...

Do you have a guess?

If you guessed, "The man was running because he was late for a very important checkers match and because he was late, the kid got to eat all of his tacos..."

You might be right! Or maybe not.

Predicting the future is hard!

Materials needed:

For our 'predicting the future' experiment, you just need a piece of paper and a pen, crayon, or marker. That's it!

What is the experiment?

On your paper, you are going to draw a big square that takes up most of the paper. Inside this square, you are going to draw three vertical lines, spaced equally apart. Then you are going to draw three lines horizontally, also equally spaced apart.

When you finish, you should have a big square on your paper with 16 smaller squares in it: four spaces across by four spaces down. You have completed what is called a "grid," which is a series of vertical lines and perpendicular lines that intersect (cross each other) and form a series of smaller, same-sized squares within the big square.

Now that you have done that, starting in the upper most left hand corner square, write the number "1," then in the next square to the right, write "2," then "3," then "4." So the top row of squares will have the numbers 1 - 4 in them. In the next row down, write the numbers 5, 6, 7, 8, one number per square. Then in the row below that, fill in those squares with the numbers 9, 10, 11, 12. And the row below that, numbers 13, 14, 15, 16.

So your numbers in your grid should look like this:

1 2 3 4

5 6 7 8

9 10 11 12

13 14 15 16

Now, above the grid, write the number "34." That will be important later!

I'm going to write about the experiment so it will really help you if you do it on your piece of paper right along with me.

Super Science

Here's the experiment: you are going to circle any number that you like. Let's say you decide to circle number 11. After you circle a number, it will look like this:

1 2 3 4

5 6 7 8

9 10 ⑪ 12

13 14 15 16

Now, every time you do this experiment, you will start off by circling one number, then you are going to draw a straight line through every number across from that number. No diagonal lines, only lines that are straight across. You'll also draw a line through every number above or below the number you chose. Remember, you won't be drawing a line numbers diagonally from the number you circled, only numbers that are above, below, or across numbers. Look at the grid again:

1 2 ~~3 4~~

5 6 ~~7 8~~

~~9 10~~ ⑪ ~~12~~

13 14 ~~15~~ 16

If you follow the rules that I just set up, that means you would have to draw a line through numbers 9, 10, and 12. Then draw lines through numbers 3, 7, and 15

Cris Johnson

Once a line is drawn through a number, you can't pick it. So if you follow that rule, the next number you could pick from those that are left would be 1, 2, 4, 5, 6, 8, 13, 14, or 16.

Let's pretend that you decide to circle number 2. So your grid would look like this:

1 ②~~3~~ 4

5 6 ~~7~~ 8

~~9 10~~ ⑪ ~~12~~

13 14 ~~15~~ 16

When you see that I circled a number, you can just circle the same number on your piece of paper.

Then you would draw a line through any numbers across from, above, or below the number 2. Remember, for this game, once a number has a line through it, you can't pick it the next time you pick a number.

Look at the grid again:

~~1~~ ②~~3 4~~

5 6 ~~7~~ 8

~~9 10~~ ⑪ ~~12~~

13 ~~14 15~~ 16

Super Science

That means now the numbers 1, 3, 4, 6, 7, 9, 10, 12, 14, and 15 are all eliminated. It's time to pick another number! From the numbers that are left, let's pretend you pick number 5, so the circled numbers in your grid would look like this:

~~1~~ ② ~~3~~ ~~4~~

⑤ ~~6~~ ~~7~~ ~~8~~

~~9~~ ~~10~~ ⑪ ~~12~~

~~13~~ ~~14~~ ~~15~~ 16

If you draw a line through all the numbers across from, below, or above 5, you have now eliminated numbers 8, and 13. All of the other numbers across, below, or above the number 5 have already been eliminated. That means only the number 16 remains, so you would circle that as your only choice, like this.

~~1~~ ② ~~3~~ ~~4~~

⑤ ~~6~~ ~~7~~ ~~8~~

~~9~~ ~~10~~ ⑪ ~~12~~

~~13~~ ~~14~~ ~~15~~ ⑯

So now, every number on the grid has either been circled or it has a line drawn through it.

The numbers that are circled are: 2, 5, 11, and 16.

Add up the numbers that are circled. I'll add the numbers up starting with the smallest numbers and work my way up to the biggest numbers:

2 + **5** = 7

7 + **11** = 18

18 + **16** = 34

Remember the number 34?

How did I know that the numbers circled would add up to 34?

I did something amazing! I predicted the future!

"Wait a minute, Mr. Science Guy," you might be thinking. "You wrote this stuff in your book. You decided what numbers to circle. You could have figured out what four numbers would add up to 34! Are you just trying to trick me?"

If you are thinking that, good for you! It means you are thinking like a scientist. You are not just agreeing with everything. Instead you are asking questions and trying to get answers!

Super Science

Here's what you can do to see if I am right. Get another piece of paper and draw yourself another grid. Then fill in the numbers, so you have another grid that looks like this:

1 2 3 4

5 6 7 8

9 10 11 12

13 14 15 16

Now go do the same experiment again. Remember, these are the rules:

- Start off by selecting any of the 16 numbers. Circle any number.
- Once you have circled a number, draw a line through any other numbers across from, below, or above, the number you circled.
- Once a number has been circled, you can't pick it again.
- Once a number has a line through it, it is eliminated and can't be circled.
- From the numbers that remain, pick another number and circle it.
- Draw a line through any numbers across from, below, and above, that number. You might wind up drawing a line through a number that already has a line through it, and that is okay.
- Keep going until every number on the grid either has a line drawn through it OR it has been circled.

- Once every number has been used, add up the numbers that have been circled.
- Helpful Tip: Every time I do this experiment, I wind up having 4 or 5 numbers that are circled.

I predict that your total will also be 34!

Once you finish doing the experiment, come back to this book.

I'll wait.

Was your total 34?

Isn't that COOL? Now you know how to do a math experiment where it looks like you can predict the future!

When I do this experiment in live school assembly shows in schools like yours, I use a great big dry erase board so hundreds of kids can see the experiment at once. Then I have different kids in the audience pick the numbers that are circled.

Before the show, I have given a teacher a big envelope that has the number 34 written really large on a piece of paper. Then I tell the audience I have predicted the future and when the envelope is opened and the audience sees the number 34, they all go, "WOW!"

You can do the same thing with your family and friends! If you follow the steps as I wrote them, your total will always be 34, no matter what numbers you decide to circle.

Super Science

It's fun!

Why is it a "superhero power?"

This is a superhero power because no one has the ability to look into the future and say, "this will definitely happen." The future is not here yet. It has not happened, so there is no way to know for sure what will happen tomorrow, or next year.

Here's the thing, though. If you look at stuff that has happened before, then you can make some really good guesses about what is probably going to happen tomorrow or even next year.

Let's say that every single day, your mom or your dad makes themselves scrambled eggs for breakfast. If you see them come home from the grocery store with a carton of eggs, you might say to yourself, "I'll predict that Mom (or Dad) is going to have scrambled eggs for breakfast tomorrow."

What happens tomorrow? You might be right...especially if your mom or dad always has scrambled eggs for breakfast.

But maybe tomorrow comes, and let's pretend there was a bad thunderstorm late at night, and the power in your home was out for a while, and then when the power comes back on, the alarm clock is doing that annoying blinky thing. That means your mom or dad's alarm is no longer set, so they might sleep too late.

When they wake up late, they might realize they slept too late and say, "Oh no! I slept too late! I don't have time to make my scrambled eggs! I guess I'll just break the eggs into a cup and drink them with my coffee. That will save me time."

So, even if your mom or dad always has scrambled eggs for breakfast, if you predict that they will definitely have scrambled eggs for breakfast the next day, something could happen to change their mind and mess up your prediction.

Later on in this chapter, I'll write about ways scientists use things to try their best to predict the future, even if they don't always get it right.

Where should I do this?

This is an experiment you definitely want to do inside…well, I suppose you can do it outside, but if it's a windy day, your paper might get blown away!

Safety:

I can't think of any way you could hurt yourself doing this experiment other than the dreaded paper cut! So don't drag the edge of the paper across your skin. And don't poke yourself with the pointy end of the pen, pencil, crayon, or marker that you are using.

Super Science

What are some others ways to experiment with this:

I have not tried this yet, but I wonder what would happen if you put different numbers in the grid, like this:

5 6 7 8

9 10 11 12

13 14 15 16

17 18 19 20

So the numbers are still in order, but I skipped number 1, 2, 3, and 4. Then I added 17, 18, 19, and 20. Try the experiment again and see if you keep coming up with the same answer? It probably will not be 34 because we changed the numbers, but maybe a different number will be the answer every time - like maybe whenever you add up the circled numbers, it will be 56, or something like that.

See what happens!

Then you can try a bigger grid - like instead of 16 squares, try a bigger grid like this:

1 2 3 4 5

6 7 8 9 10

11 12 13 14 15

16 17 18 19 20

If you do the experiment again, with the same rules, try it a few different times and circle different numbers every time to see if adding the circled numbers gives you the same answer every time.

The science of the experiment:

This experiment uses something called an "algorithm." That's a word that means you follow a step by step process in order to get something done that you want to get done.

You see things like this all the time in life. For instance, how about a recipe to make a cake? There are certain steps that a good cook follows and when they do all the steps correctly, then they make a yummy cake!

If they make a mistake, then maybe the cake doesn't turn out quite so yummy. Let's pretend a cook is trying to make a yummy chocolate cake. The recipe calls for a cup of sugar, but instead the cook puts a cup of mustard into the recipe! What do you think will happen? Yuck!

Once a person learns how to do something, they have an algorithm, or recipe, to make sure they don't miss a step. They do this so every time they do that process, they get the same result.

One time, my wife and I went to one of our favorite restaurants. We both ordered one of our favorite dishes. Every time we went to the restaurant and ordered this

Super Science

dish, it was always the same and we looked forward to it being the same.

The last time we went to this restaurant, we ordered our favorite dish…and when the food came, there were a bunch of the ingredients that had been changed. It was gross!

We asked the people at the restaurant why the dish had been changed, and they said, "Oh, the cook ran out of some of the ingredients. So he just tossed in other stuff he had laying around."

Wow! That was terrible. We never ordered that dish or went to that restaurant again.

I wish I could tell you the name of the restaurant but since this is in a book, I can't. Ask your parents or your teacher what "litigation" means!

Similar experiments:

There are a lot of really cool experiments like this involving grids and numbers. One of the more well-known experiments is called a "Magic Square."

This is pretty advanced math, but if you are reading this book, then it means you like science, you like figuring stuff like this out, and you will be able to understand this stuff if you take your time.

Cris Johnson

In this experiment, a scientist draws a grid just like the one we have done in this experiment. The difference is with this grid, no numbers are filled in yet.

Then the scientist asks someone watching to give her a two-digit number...we will pretend that the number is 35. Then the scientist starts filling in the other squares with other numbers. The numbers she fills in are not in order and none of the numbers is 35.

The scientist then starts adding up the numbers in the first row. All for numbers in that first row add up to 35! Then she adds up the second row of four numbers. They also add up to 35! Then the third row: 35! Fourth row: 35!

Next the scientist adds up the first column of four numbers. That is the left vertical four numbers. 35! All the columns add up to 35! Then she draws a diagonal line from the top right corner to the lower left corner. She adds up those numbers. 35! Then she draws another diagonal line, this time from the upper left corner to the lower right corner. 35!

It's amazing!

There are a few ways to do a Magic Square. The first way is by getting REALLY good at math so that when someone gives you a number, you can work out all of the math in your brain in just a few minutes and then write in the numbers. This takes a LOT of work, but if you practice enough and study math enough, you can do it!

Super Science

The other way is with another algorithm, or recipe. In this case, some of the numbers are the same in the grid every time. Here is one I have used. I made four of the squares **bold, like this:**

11 14 5 **x-30**

4 **+1** 10 15

+2 7 12 9

13 8 **+3** 6

The numbers you see that are NOT **bold** remain the same every time. But the squares that say "x," "+1," "+2" and "+3" are numbers that change depending on what number someone gives you.

The Magic Square I just wrote will work for any number larger than 30. Let's pretend that someone gives you the number 35.

The algorithm for this to work is very simple. For those special four squares, you would take the number someone gives you and do some math that will give you four new numbers to put in those four places in the grid. Here are the math problems to do:

Your number minus 30.

Your number minus 30 plus 1.

Your number minus 30 plus 2.

Cris Johnson

Your number minus 30 plus 3.

So if the number someone gives you is 35, the answers would be:

35 - 30 = 5

35 - 30 + 1 = 6

35 - 30 + 2 = 7

35 - 30 + 3 = 8

Now take those four answers and put them in the right places in the grid I made them **bold, like this:**

11 14 5 **5**

4 **6** 10 15

7 7 12 9

13 **8** 8 6

Now, add up the top row:

11 + 14 = 25

25 + 5 = 30

30 + 5 = 35

See? It works! It seems pretty amazing. You can add up the rest of the rows and they WILL each add up to 35. You can do the same thing with the columns, and each will add up to 35. Try the diagonals: They will each add up to 35, too!

Super Science

This is an algorithm, or a recipe, for getting to a result you want. It's not a superhero power but it IS really cool!

If you want to have fun with this, try it with different numbers, any number larger than 30. Because those four special squares are going to change every time you do the experiment, you will need a new piece of paper every time. Or if you have a whiteboard or something like that, you can just erase the four squares every time.

You'll want to check your math by adding up the four rows, then the four columns, then the diagonals. Every time you add up four of the numbers, the answer should be the original number you picked each time!

If you get a different number, then it means you made a mistake with your addition somewhere. If you do make a mistake, it's really good exercise for your brain to go back and figure out where the mistake is.

Which of the 4 sciences does this fit:

This is definitely an experiment in the Math Sciences! At it's most basic, there is a lot of addition involved because we are adding up numbers to get to that result that we are expecting.

Math is SO important because we use it in so many ways. Remember the recipe I wrote about earlier? That uses math to make sure food comes out the same every time.

Scientists like having math algorithms that give them the same result every time! Some scientists use really complicated math to try to predict things that are really hard to predict...like the weather! A meteorologist is a scientist who studies weather patterns and tries their best to figure out what is going to happen with weather so people can plan what they are going to do and when.

The website where you can see the experiments:

www.SuperheroScienceBook.com

You will need your parents to fill out the contact form to get access to the videos.

Make sure you only go on the Internet with your parent's permission!

Our next chapter is our last experiment and I saved this for last because it is one if my favorite experiments ever, and it's the experiment that gave me the idea for this entire book!

Cool Science Questions:

1. What is an algorithm?
2. How is a recipe like an algorithm?

Super Science

Chapter 14: Controlling the Elements

This science experiment is the entire reason I became interested in science! I love this experiment because:

- It's really fun and messy
- It's not very complicated
- it's pretty easy to do
- It's safe
- It looks amazing!

Let's get into it!

What is the superhero power?

There are some superheroes who have the ability to "control the elements." The 'elements' are fire, water, air, and earth. So that means to control the elements, the superhero (or super bad guy) would be able to cause wind storms, create or move fire, lift big rocks with the power of the mind...or control water.

Cris Johnson

Most superheroes who have this power simply use the power of their mind, which means that when you see them doing this, they are squinting really hard to show they are thinking about it very carefully.

Other superheroes might use special machines to control the weather, but either way, it's an ability that is incredible to see in comic books, superhero movies, and TV shows.

It's an amazing superhero power, and one that could be dangerous if we were careless when trying to recreate it with science. So we are going to stay away from the fire, big rocks, and the air.

For this experiment, we are going to make it look like we can cause water to launch up into the air 20 feet or more, all by itself!

Materials needed:

For this super-fun experiment, you'll need just a few things:

- A 2-liter bottle of Diet Coke
- A package of original Mentos candies
- A piece of card stock, such as a 4"x6" notecard
- Scotch tape
- A flat level surface to do the experiment
- Safety goggles

Super Science

I know I just wrote that you need to have Diet Coke, but really, I have done it with generic diet cola to save some money. This experiment is known as "Diet Coke & Mentos" as that's what most people doing the experiment used, but you can use a cheaper cola and it still works great.

While any kind of soda should work for this experiment, most scientists who do this use some form of 'diet' cola because diet colas are less sticky. Since we are going to be making a huge mess with this, using something that is less sticky might be important!

For the candies, it has to be Mentos, the original kind that feels bumpy when you touch it. Later on, I will explain why that is important to the experiment. I will also include some other things you might be able to try, just for fun.

For the cardstock, if you or your parents do not have index cards, you can also just use a piece of very thin cardboard. As long as it's thin enough to easily roll in a tube, it will work. A regular piece of paper would work too, but I like using cardstock because it's thick enough to not get all 'bendy' when I'm rolling it in a tube.

The safety goggles are there to make sure you don't get any soda in your eyes because that would be super-gross!

What is the experiment?

Start off by taking the cardstock (or piece of paper or whatever you are using) and roll it into a thin tube. You are going to hold this tube up so if you drop a candy into the opening on the top, it will fall right through and come out the bottom.

The tube has to be thin enough to fit into the opening of the bottle of soda.

The opening in the tube also must be big enough for those candies to fall right through without getting stuck.

Once you have your tube rolled into the right size, tape it place with the Scotch tape.

Put the bottle of diet cola on your level surface, such as a table or driveway. DON'T shake up the soda, because when you open the bottle, you don't want soda to spill out...that's coming later!

Open the bottle of soda.

If you buy a big bag of Mentos candies, each candy is wrapped separately. You will want to unwrap 4, 5, or 6 candies so they can be dropped into the bottle of soda all at once.

Really, you can drop fewer than 4 candies, but that will give a lower reaction and where is the fun in that? You can also do more than 6 candies too. I have done it with as many as 8 candies at once, so you can choose!

Make sure your safety goggles are on to prevent getting sticky cola in your eyes!

Once you have your bottle on the level surface and your candies are unwrapped, use one of your hands to block the bottom of the tube. I usually use my thumb on my left hand. Then drop the candies into the tube with your other hand. All the candies will be held in place in the tube.

Hold the candy-filled tube right over the opening of the soda bottle. Take away your fingers that are blocking the bottom of the tube. This will allow the candies to fall right into the soda bottle.

Then, stand back! Get away from the bottle as fast as you can!

About 2 seconds after you drop the candies into the bottle, a huge geyser of soda is going to come rushing up out of the bottle! If you used 6 or more candies, the soda would launch into the air about 20 feet! This is amazing to see!

It might not be the superhero power of controlling water, but it's still really cool!

Why is it a "superhero power?"

As I wrote earlier, controlling the elements is considered a superhero power. Let's look at the four elements, earth, wind, water, fire.

People do not have the ability to control fire. In fact, when people on the set of a movie use real fire, they take all kinds of precautions because fire can get out of control really quickly and dangerously and the slightest mistake can mean people can get hurt.

People also do not have the ability to control earth. If there is an earthquake, there really isn't anything anyone can do except hang on and wait for it to be over. About the most people can do to 'control' the earth is get big machines to scoop earth, remove big rocks, and things like that. But that's not controlling anything...it's just landscaping!

Wind? We can't control the wind blowing around us. The best we can do is build stronger buildings to protect us from super-strong winds such as tornadoes or hurricanes, but even then, Mother Nature is usually strong enough to blow over anything humans can build.

Now we get to water. We can't control water...but we can kind of, sort of, guide it or direct it sometimes. If you have ever seen a dam, then you might have an idea of what I mean. A dam is something that is built to block water or change it's direction.

People build dams to make ponds or lakes. And some animals, like beavers, do the same thing! Beavers are so cool.

Where should I do this?

You can do this in your bedroom!

No! I'm just kidding! PLEASE don't do this in your bedroom, or anywhere indoors!

Super Science

This experiment is SUPER fun but also so completely messy that if you try to do this indoors, I promise your mom or dad will take this book away and never let you do science experiments ever again.

So, this HAS to be done outside. And when you do this outside, it should be in a place where it won't matter if you get sticky soda everywhere. That means you need to pick a place that is not too close to your parents' car, or the building you live in, or anything that shouldn't get sticky.

Oh, if you have a dog or a cat, make sure they are not anywhere near this experiment because trying to clean up a sticky dog or a cat is just no fun. No fun at all.

Safety:

I think as far as safety goes, this experiment mostly just makes a big mess so there really is not much that can hurt you, other than your eyes getting soda in them, so always wear those safety goggles!

What are some others ways to experiment with this:

I filmed myself doing every experiment in this book. You have probably read this in other places in this book, but I know some readers will skip around different parts of this book. On the video that I did based on this experiment, I first put 6 candies into the soda and that gave me a really cool, really tall geyser!

You can try 8, 10, 15 candies or more! See what happens!

I also filmed myself doing the experiment with just one candy. Do you think the result was the same or different?

You can also try different brands of drinks. Maybe some sodas make a better geyser. Maybe regular cola is better than diet. These are all things you can try.

Have contests with your friends! See who has the highest geyser!

You can also try different candies if you can find candies that have the same kind of little bumps and little holes on them. You can also try rock salt, which is a lot bigger than table salt. It also has all kinds of bumps and holes in each piece.

Guess what? All those little bumps and holes in the candies are a big reason why this experiment works!

The science of the experiment:

There are a couple of things that work together to make this work. The first is the soda itself - it's carbonated. "Carbonation" is the gas carbon dioxide dissolving in a liquid. When the soda is packed at the factory, that carbon dioxide is held in place by the pressure of the sealed bottle. This is what makes the soda "fizzy" as you drink it.

When the bottle is opened, the pressure is released and that carbon dioxide dissolves, those gas bubbles want to get out of the liquid. If you pour a drink of soda in a clear gas, you can watch the tiny bubbles as they scoot up to the surface of the liquid. Many of those bubbles are trapped by the "surface tension" of the liquid. Surface tension means the liquid molecules are attracted to each other and they stay together strong enough to prevent many of the bubbles from popping - they rise to the top and form foam.

So that's the first part. The second part involves the candies themselves.

If you throw a sold object into a glass of soda, you'll see many little bubbles form on whatever you put in the glass. That is the carbon dioxide trying to escape.

Mentos are very special candies because they have thousands of little teeny weeny cavities on them. That's why they feel bumpy when you touch them. When those candies are tossed into the soda, the huge number of those cavities allow the carbon dioxide gas to form bubbles on the surface of each candy.

But that's not all! Because the candies are heavy enough to sink to the bottom of the bottle, the reaction is even more intense and fast, meaning the liquid shoots out of the bottle in a huge geyser!

Similar experiments:

A very popular experiment that also involves carbon dioxide is the Baking Soda Volcano! In this experiment, baking soda is mixed with vinegar and carbon dioxide gas is formed and released, which looks like a volcano erupting! This is a classic experiment that has been around for a very long time.

Another carbon dioxide experiment that I will sometimes do in my live assembly programs when I visit schools is using baking soda and vinegar to blow up a balloon!

This is very easy to do. You'll need:

- an empty water bottle
- vinegar
- baking soda
- a 9 inch round balloon

To do the experiment, put about a tablespoon of baking soda into the balloon. You might need a small funnel to do this.

Next, pour some vinegar into the empty water bottle. Just pour enough vinegar in the bottle so it's about two inches deep.

Then, you need to wrap the nozzle, or opening, of the balloon around the opening of the bottle. Just stretch it and wrap it around. You want to make sure none of the baking soda gets into the vinegar yet.

Once the balloon is securely around the opening of the bottle, lift up the main body of the balloon where the baking soda is and shake, wiggle, and bounce the body of the balloon to get the baking soda powder to drop into the bottle and mix with the vinegar.

As soon the baking soda is in the vinegar, shake the bottle a little to mix the powder with the liquid. In a few seconds, carbon dioxide is released and the gas will fill up the balloon! This looks really cool!

Which of the 4 sciences does this fit:

This is a Physical Sciences experiment. What is interesting is that some scientists once thought this was a Chemistry experiment. They thought there was something in the candies that actually mixed with the soda and caused the reaction.

That is not the case, because it is how the candy Mentos is made that causes this reaction to happen in such a spectacular way! In fact, if you were to take another candy that was the exact same shape and made with the exact same ingredients, but the surface was smooth, the reaction would not happen.

The website where you can see the experiments:

www.SuperheroScienceBook.com

You will need your parents to fill out the contact form to get access to the videos. Make sure you only go on the Internet with your parent's permission!

Cris Johnson

Cool Science Questions:

1. Is the Diet Coke & Mentos a chemical reaction or physical reaction?
2. What is carbonation?
3. What is surface tension?

Super Science

Chapter 15: Conclusion

I hope you have enjoyed this book! I have always wanted to write a book ever since I was the age of most of you reading this book!

Writing this book has been a lot of fun for me and it reminded me of a lot of science experiments I enjoyed years ago.

For those of you reading this, I really hope this book gets you excited about science enough to actually try some of the experiments in this book. I picked experiments that were safe, fun, and inexpensive.

Always remember to stay safe when doing science experiments and have a grownup helping you just in case something goes wrong.

I'm also working on ideas for two more science books. One of them will be another superhero-themed science book, while the other will be about simple machines, so when

Cris Johnson

you ask your parents to sign up to see the Super Science videos, they will get a note sent to them when my next book is ready.

Also, maybe in my next science book, we will answer the question of why the hot dog was waiting for that bus!

Super Science

BONUS
Chapter for
Parents & Teachers

Parents & Teachers,

Thank you for reading this! After delivering live school assembly programs to kids in 200 schools each year in multiple states for the last 20 years, I have always wanted to reach an even broader audience. This book is the result of that desire.

The vast majority of my live assemblies are presented to elementary school kids, so I wrote the book with those ages in mind.

My purpose in writing this bonus chapter was to compile a list of most of the terms, concepts, and experiments in this book as a way to illustrate that through my admittedly goofy writing style, a lot of sound educational concepts are being presented.

If you have not read the entire book, I also provided some review questions at the end of each chapter so if they were so inclined, kids reading this book could test how much information they retained.

I will list each chapter and then provide a summary of the content within each.

So let's dig in!

Chapter 1: Introduction: Science & Technology

In this chapter, I established what the premise of the book is, namely learning about science experiments that sort of resemble superhero powers. I also reiterate my goal to give kids fun experiments that they can easily do themselves without having to buy any expensive equipment.

I also outlined how the experiments would be presented along with various subtopics in each chapter, such as safety, materials needed, and so forth.

Near the end of the chapter, I also linked science with technology and explained that without science, kids would not have access to all of their cool smart phones, games, and such.

Super Science

Chapter 2: The Scientific Method: Flying Hot Dogs & Buses

My goal with this chapter was to not only illustrate the steps in the Scientific Method in ways that kids could easily understand but to do so in extremely goofy ways so as to maintain the interest in young readers.

I did so by posing the question: "Can hot dogs fly?" This seemingly ridiculous premise establishes something that kids already know - hot dogs cannot fly - as a way to illustrate the Scientific Method without bogging down the explanation by simultaneously trying to explain something that perhaps we *don't* know.

On top of that, I aimed to introduce kids to the concept of **critical thinking** in that they need not make assumptions but follow through and be sure, as much as they can, of their conclusions.

I also spent some additional time illustrating why a **hypothesis** is more than just a guess, by explaining how the idea of the President of The United States being named as the culprit in the theft of a kid's bike is not a valid hypothesis, thereby introducing the concept of an **educated guess.**

Admittedly, this chapter was the most fun for me to write!

Cris Johnson

Chapter 3: The Four Sciences

Here the goal was to introduce readers to the four "main" categories of science:

Physical Sciences

Life Sciences

Math Sciences

Social Sciences

From there, I dived a little more deeply into each, introduced some specific areas of study in each, along with the names of some of these specialties.

Under Life Sciences, I defined **veterinarians, ornithologists**, and **pediatricians** simply to give readers an idea of how much variety of sciences there are in just one category.

For Physical Sciences, **electrostatics, meteorology**, and **meteorologist** were defined.

I took a slightly different route with Math Sciences and spent some time establishing how the study of **Mathematics** was applicable to studying so much more than just algebra and geometry and really could apply to the universe.

I took this approach to help kids appreciate that math was so much more useful than perhaps they thought. I myself had a hard time believing that I would "use any of this math stuff in the real world" after I graduated, so I reasoned other kids may feel the same way.

I illustrated the Social Sciences primarily by writing about studying human behavior, explaining a little bit about **stress, phobias**, and **optical illusions** and talking about my other job, as a hypnotist, where I help people change their thoughts about certain things so they feel better.

Chapter 4: X-Ray Vision

The main thrust of this chapter was teaching kids the classic 'hole through your hand' optical illusion. Using the framing device of this mimicking a superhero's ability to see through solid objects, this can be quite exciting to kids!

Along the way, I touched upon how some technology allows people to see or view things through solid objects. I briefly discussed different uses of **x-rays** and **thermal imaging.**

For the experiment itself, **binocular vision** and **psychology** were terms explained as well.

Chapter 5: Creating Solid Matter Out of Thin Air

The Elephant Toothpaste experiment is the focus of this chapter! This was enjoyable to me as until I began working on this project, I had never tried this experiment, so I learned a lot.

The concept of *'what is matter?'* was explored, along with *states of matter*, and specifically *liquids, solids,* and *gases* were defined.

Along the way, *yeast* was defined and examples given where yeast might be used and why.

Other concepts defined and/or illustrated include: *chemistry, catalyst, quadrant, concentrated, exothermic reactions*, and *diluted*.

Chapter 6: Telekinesis: Stopping A Falling Object

My goal with this chapter was to find a simple science experiment to illustrate one of the most impressive superhero powers, that of invulnerability. By doing the simple yet fun "Leakless Bag" experiment, kids are introduced to basic concepts of gravity as well as having *telekinesis* defined.

Along the way, I reveal the true story of how a magician named Steve Shaw once fooled several scientists into thinking he had true telekinetic ability. The purpose of this is to impart upon readers that true science experimentation involves no bias, no goals other than finding out the truth.

I also touch upon *Isaac Newton* and briefly cover just one of the reasons why he is so important to science.

Other terms defined include: **parallel, turbulence,** and **gravitation**.

Chapter 7: Telekinesis: Make Objects Fly Into the Air

Telekinesis is once again the focus of a 'superhero power' in this chapter but presented in an even more dramatic way.

The "Stacked Ball Drop" experiment was used to illustrate concepts of **potential energy, kinetic energy,** and **elastic potential energy.**

Additional terms defined: **physics** and **essential**.

Chapter 8: Mutations

This chapter was again used to illustrate concepts of **optical illusions** by using the classic "Third Finger Illusion," which has been around at least 100 years. The superhero power, as illustrated by the chapter title, was the idea of mutations and growing extra body parts which would, in comics, give those mutants all kinds of cool abilities.

I also discussed what humans can and *can't* regenerate when injured.

Additionally, when discussing the superhero power of mutations, an offshoot of that ability is if the hero has the ability to generate multiple versions of him/herself, or at least additional images. From there, I included a brief discussion of hypnotic phenomenon of hallucinating when in certain states of mind. Fascinating stuff!

Chapter 9: Creating Evil Serpents

Focusing this time on what kind of evil creature a 'mad scientist' might create, this chapter introduced readers to the fun, safe "Bubble Snakes" experiment!

The structure of bubbles were discussed, including why a sphere was the most efficient, practical shape for a bubble, as well as practical studies going on in labs involving bubbles!

Since the comic book aspect of creating creatures was the focal point of the chapter, I included a discussion on **cloning**, and some current thinking in the scientific community about whether or not we will ever see cloned humans.

I also stress, as I have multiple times in the book, the need for **safety**. I illustrate this by explaining that even grown ups can have what looks like very silly accidents by recounting a story about me cutting myself badly with a pair of scissors performing a magic trick!

Additional terms and concepts discussed include: **oxygen, carbon dioxide, circumference, engineering,** and **weightless environment.**

Super Science

Chapter 10: Passing Through Solid Objects

The superhero power of passing through solid objects opened the door to a really amazing optical illusion where it looks like one of your hands is passing through the other!

The first time I saw this, it completely floored me and I was sure I was watching a CGI effect of some kind. However, as kids reading the book will discover, one of the main concepts that makes this illusion so convincing is that of **conditioning**.

This led to a discussion on **psychology, Ivan Pavlov**, and his experiments with dogs and conditioning.

Additionally, I also discussed the law of physics stating that two objects cannot occupy the same space at the same time.

Chapter 11: Invulnerability

Being able to withstand damage without getting hurt is a popular superhero power and one that led to the fun "Leakless Bag" experiment. Branching off from this starting point, I briefly covered some of the ways animals have evolved different forms of protection for themselves.

The big discussion point in this simple yet fun experiment is **polymers**, and their myriad of uses in our daily lives with different products.

Other terms and concepts discussed include: **Chemistry**, chemistry scientist and Nobel Prize recipient **Hermann Staudinger, natural rubber,** and **Charles Goodyear.**

Chapter 12: Shrinking

Forced Perspective is the name of the game in this experiment that seemingly replicates the superhero power of shrinking down to "action figure size."

Along the way, I discuss that while people can't shrink down to be super tiny, as humans get older they do lose a bit of height.

What I really love about this experiment is that it's a nice change of pace from everything else in the book in that this experiment is more about creativity than anything else.

Other terms and concepts discussed: **muscle atrophy, perceptions, optical illusions,** and **psychology.**

Chapter 13: Predicting the Future

Who wouldn't want to be able to predict the future? Whether it's winning the lottery or simply knowing what's going to happen in your own life, this is a very popular superhero power for a science book.

This experiment was a simple math trick I learned from a magic book. Super simple and it fooled me completely when I saw it performed the first time.

Super Science

What I enjoyed most about writing this chapter was explaining to the kids the fact that from a certain point of view, we can predict our future...by using **algorithms**, formulas, and recipes. These concepts allow us to get a certain result, whether it is a math problem, a Magic Square, or even a cake recipe.

Chapter 14: Controlling the Elements

This, more than anything else in this book, was what motivated me to put together **Super Science**, both this book as well as my virtual school assembly program by the same name.

The experiment, "Diet Coke & Mentos," was something I first learned about back in 2006 from the Discovery Channel show, *Mythbusters*.

I framed this crazy-looking experiment around the idea of superheroes being able to control elements, such as water, fire, wind, and earth. Pretty powerful superhero power! This experiment is certainly more mundane than that fictional comic book idea, but I have never met a kid yet who didn't think the Diet Coke & Mentos experiment was awesome!

Terms and concepts discussed include: **carbonation** and **surface tension**. I also touch ever-so-lightly on the concept of **nucleation**, even though it is not specifically named. This was one experiment where I really tried to simplify some of the more complicated science ideas.

Cris Johnson

Chapter 15: Conclusion

This chapter was a simple wrap-up for the kids along with another reminder to use safety when doing scientific experiments.

Writing this book has been a lifelong dream of mine! If you have any comments or questions, please let me know.

Feel free to email me at cris@elementaryschoolassemblies.com.

Also, don't forget to access the free performance videos simply by visiting:

www.SuperheroScienceBook.com

Simply enter your information in the contact form on that webpage and from there you will be emailed the URL to gain access to videos of me conducting and explaining each experiment.

I am also hard at work on two new books aimed at young readers. When you sign up for access to the Super Science videos, you'll be added to an email list and I will notify you when my additional books are complete.

Thank you once again for your attention!

Warmly,

Cris Johnson

Super Science

See Cris Johnson LIVE!

Cris Johnson offers educational school assembly programs for students in select states across the United States and motivational programs for school faculty!

Visit www.ElementarySchoolAssemblies.com for dates of availability and program information.

www.ingramcontent.com/pod-product-compliance
Lightning Source LLC
Chambersburg PA
CBHW030906080526
44589CB00010B/178